Urban Air Pollution Monitoring by Ground-Based Stations and Satellite Data

Mikalai Filonchyk • Haowen Yan

Urban Air Pollution Monitoring by Ground-Based Stations and Satellite Data

Multi-season Characteristics from Lanzhou City, China

Springer

Mikalai Filonchyk
Department of Geographic Information Science
Faculty of Geomatics
Lanzhou Jiaotong University
Lanzhou, China

Haowen Yan
Department of Geographic Information Science
Faculty of Geomatics
Lanzhou Jiaotong University
Lanzhou, China

ISBN 978-3-319-78044-3 ISBN 978-3-319-78045-0 (eBook)
https://doi.org/10.1007/978-3-319-78045-0

Library of Congress Control Number: 2018937676

Printed on acid-free paper

This Springer imprint is published by the registered company Springer International Publishing AG part of Springer Nature.
The registered company address is: Gewerbestrasse 11, 6330 Cham, Switzerland

Preface

In China the past 20 years have been characterized by rapid economic growth and urbanization. The concentration and intensification of human activities are so significant that the natural environment has ceased to conform to the normal conditions of human life.

Modern Chinese cities are powerful systems, and every city has changed, both within its own territory and far beyond. At the same time, the atmosphere, the geographical environment of Earth that is the most mobile and sensitive to contamination, has been subjected to significant human impact. High concentrations of pollutants in the atmosphere of the cities and industrial centers causes great harm to public health, as well as damage to agriculture, industry, residential buildings, technical facilities, and historical monuments. The fact that regional environmental problems have become particularly acute in China in recent years is no coincidence.

Lanzhou City, in Northwestern China, is currently facing a serious air quality problem, with key factors such as particulate matter in the air leading to serious and increased urban pollution. Thus, it is vital to control the pollution level and the various atmospheric pollutants in the city and to evaluate its air quality; accordingly, reasonable prediction methods must be found to control its air quality. Therefore, in this book we aimed to study the role of natural and anthropogenic factors in the formation of air pollution in Lanzhou City, the main industrial center of Northwestern China. We used the following methods: mathematical statistics, the hybrid single particle Lagrangian integrated trajectory model (HYSPLIT), and Cloud-Aerosol Lidar and the Infrared Pathfinder Satellite Observations (CALIPSO), as well as the ozone monitoring instrument (OMI) method of predicting aerosol pollution by the Navy Aerosol Analysis and Prediction System (NAAPS), utilizing the data obtained from ground-based air quality monitoring stations and the MODIS (Terra) and Landsat-8 satellites. The study is based on data obtained from January 1, 2003, to December 31, 2016, covering different seasons.

The authors would like to express their appreciation to many people who made the completion of this book possible. Above all, the first author is grateful to Professor Yang Shuwen in the Faculty of Geomatics, Lanzhou Jiaotong University, China, and Professor Zhao Baowei in the Faculty of Environmental and Municipal Engineering,

Lanzhou Jiaotong University, China, for their valuable and competent advice and guidance, and for their immense support and contribution. The first author is also grateful to Associate Professor Alexander N. Kusenkov in the Department of Ecology, Gomel State University, Belarus, who indirectly helped with the completion of the book. The first author would like to express his profound gratitude to his family—especially his mother, whose help, prayers, encouragement, and kind cooperation helped throughout the entire period of this study—and to thank his friend Volha Hurynovich, for her incredible support and wonderful contribution toward the completion of this research.

Last but not least, the authors appreciate the National Key R&D Program of China (2017YFB0504203 and 2017YFB0504201), the National Natural Science Foundation of China (No. 41371435), and the Talent Innovation Project of Lanzhou (No. 2015-RC-28) for their financial support of the work described in the book.

This book can be used as a reference for graduates and researchers who are interested in ecology and air pollution, especially dust air pollution and atmospheric particulate matter $(PM)_{2.5}$ or PM_{10}. Any comments and suggestions regarding this book will be greatly welcomed and appreciated.

Lanzhou, China Mikalai Filonchyk
Lanzhou, China Haowen Yan

Contents

Abbreviations

°C	Degrees Celsius
μg/m^3	Micrograms (1-millionth of a gram) per cubic meter air
μm	Micrometer
τ	Optical depth or optical thickness (usually taken to mean "aerosol" optical depth, AOD, or optical thickness, AOT). May have subscripts to symbolize wavelength or measured by which instrument (e.g., $\tau_{0.55}$)
a.m.	Indicating the time period from midnight to midday (ante meridiem)
AAI	Absorbing aerosol index
AOD	Aerosol optical depth
AOT	Aerosol optical thickness
API	Air pollution index
AQI	Air quality index
CALIPSO	Cloud-Aerosol Lidar and Infrared Pathfinder Satellite Observations
HYSPLIT	Hybrid Single Particle Lagrangian Integrated Trajectory model
K'	Meteorological self-cleaning capacity of the atmosphere
km	Kilometer
LST	Land surface temperature
m/s	Meter per second
MEP	Ministry of Environmental Protection of the People's Republic of China
mg/m^3	Milligrams (1-thousandth of a gram) per cubic meter air
mm	Millimeter
MODIS	Moderate-Resolution Imaging Spectroradiometer
NAAPS	Navy Aerosol Analysis and Prediction System
NAAQS	National Ambient Air Quality Standard
NDVI	Normalized difference vegetation index
nm	Nanometer
NO*x*	Nitrogen oxides
OMI	Ozone Monitoring Instrument
p.m.	Indicating the time period from midday to midnight (post meridiem)
PAP	Potential atmosphere pollution

PM_{10} Concentration of particulate matter less than 10 μm in diameter
$PM_{2.5}$ Concentration of particulate matter less than 2.5 μm in diameter
r Pearson's linear correlation
R^2 Coefficient of determination
RH Relative humidity (%)
RMA Reduced major axis
r_s Spearman's rank correlation coefficient
T Temperature (°C)
UHI Urban heat island
UTC Coordinated universal time
UV Ultraviolet
V Visibility (km)
VOC Volatile organic compounds
WHO World Health Organization
WS Wind speed (m/s)

Chapter 1
Introduction

1.1 Background

Particulate matter (PM) is a widespread air pollutant, comprising a mixture of aerosols and liquid particles of the air in a suspended state. The indicators, commonly used to characterize the PM and relevant to health, include the mass concentration of particles with a diameter of less than 10 μm (PM_{10}) and less than 2.5 μm ($PM_{2.5}$) (Pope and Dockery 2006; Shiraiwa et al. 2012). $PM_{2.5}$, often called as fine suspended particles, also includes ultrafine particles with a diameter of less than 0.1 μm (Mcmurry et al. 2004; US EPA 2010; Fann and Risley 2013). Table 1.1 shows this size categorization concept. The PM with a diameter from 0.1 to 1 μm can exist in the air for many days and weeks, respectively, and also can be subjected to cross-border transport of air for long distances.

PM is a mixture of physical and chemical characteristics, which vary depending on the location. The most common chemical components of PM include sulfates, nitrates, ammonia, and other inorganic ions such as sodium, potassium, calcium, magnesium, and chloride ions, organic and elemental carbon, minerals, crust associated water, particles, metals, and polycyclic aromatic hydrocarbons (PAHs) (Schmidt-Ott and Büscher 1991; Okuyama et al. 2005; Chen and Xiao 2009). In the compositions, PM can also be found in such biological components as allergens and microorganisms.

The particles may be either directly emitted into the atmospheric air (primary PM) or formed in the atmosphere from such gaseous pollutants as sulfur dioxide, nitrogen oxides, ammonia, and methane volatile organic compounds (secondary particles) (Wang 1999).

Primarily, PM and gaseous pollutants may occur in both natural (non-anthropogenic) and artificial (man-made) sources. Anthropogenic sources include internal combustion engines (both diesel and petrol combustion engines), solid fuels (brown coal, heavy oil, coal, and biomass), burned for power generation in the domestic sector and in industry, other industrial constructions (mining minerals; cement, brick, and

M. Filonchyk, H. Yan, *Urban Air Pollution Monitoring by Ground-Based Stations and Satellite Data*, https://doi.org/10.1007/978-3-319-78045-0_1

Table 1.1 Aerosol size categorization

Fine mode ≤2.5 μm		Coarse mode 2.5–10 μm
Ultrafine mode <0.1 μm	Accumulation mode 0.1–2.5 μm	

ceramic production; and melting), and erosion pavement due to vehicular traffic and abrasion of tires and brake pads (Bench et al. 2007; Gillette et al. 2010). The main source of ammonia is agriculture. The secondary particles in the air are formed by chemical reactions of gaseous pollutants. They are the product of transformation of nitrogen oxides (NO_x) in the atmosphere, which are emitted mainly by traffic and some industrial processes. Sulfur dioxide formed from burning sulfur-containing fuel. Secondary particles are contained mainly in the fine PM. Another source of formation of PM is the resuspension of soil and dust, especially in arid regions or during the transportation through dusted places over long distances, for example, from the Sahara desert to Southern Europe (Gobbi et al. 2007; Karlsson et al. 2008; Masson et al. 2010; Mallone et al. 2011; Remoundaki et al. 2011), as well as from the Gobi Desert to Northern and Central China (Zhang et al. 1998; Ma et al. 2008; Wang et al. 2011, 2012; Filonchyk et al. 2016).

The effect of aerosols, causing weather and climate aftermath, has a twofold direction. On the one hand, there is an effect of aerosols on clouding and precipitation processes by changing the microstructure of cloud particles; on the other hand, there is an effect of aerosols on the transfer processes of solar and thermal radiation in the atmosphere, as well as on the temperature regime of the Earth's climate system. There are direct and indirect effects of aerosols. The direct effect is that aerosols scatter and absorb solar and thermal radiation and thus alter the radiation balance of the atmosphere and the underlying surface. The aerosols contain sulfates, organic carbon, and soot from burnt fuel and biomass, dusted into the atmosphere and at the Earth's surface as a result of economic activity. They have a significant direct effect on the flow of solar and thermal radiation. In Fig. 1.1, a schematic diagram shows the most known effects of aerosols on cloud properties.

Indirect climate impact of aerosols is manifested mainly in the change of radioactive properties of clouds (absorption and reflection), as well as the time of their life in the atmosphere under their influence.

Globally, aerosols are the primary regulators of solar radiation fluxes in the Earth atmosphere, after clouds (which are also aerosol formed). Aerosol layers also absorb heat (its own) radiation of the atmosphere and the underlying surface of the Earth, putting additional pressure on the energy balance of the Earth's climate system.

Several epidemiological studies have shown that both short-term and long-term exposure of particulate matters increase the risk of morbidity and mortality. It is a global nature problem. Both developing and developed countries suffer from health problems caused by PM (WHO 2006, 2015).

PM_{10} and $PM_{2.5}$ contain particles which have so small diameter that they can penetrate deep into the respiratory system. There is evidence of the effect of PM_{10} on the respiratory system; however, in terms of mortality, a more important risk

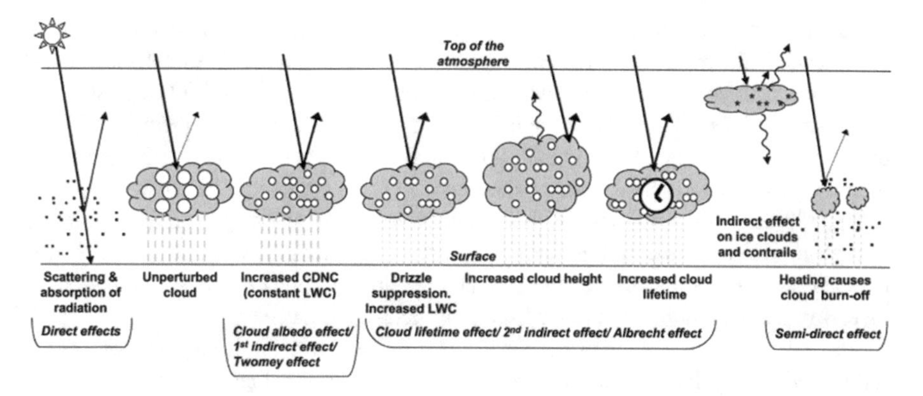

Fig. 1.1 Schematic representation of the mechanisms of action of radiation exposure on aerosols and clouds. (Small black dots represent the aerosol particles and the large open circles represent cloud droplets. Straight lines represent the incident and reflected solar radiation. The wavy line represents terrestrial radiation; filled white circles indicate the concentration of cloud droplets (CDNC). The unperturbed cloud comprises larger cloud drops, since only natural sprays are available as cloud condensation nuclei, while a large number of smaller cloud droplets are contained in the disturbed cloud, as both natural and anthropogenic aerosols are available as cloud condensation nuclei (CCN). LWC refers to the content of liquid water)

factor is the coarse fraction of PM_{10} and $PM_{2.5}$. It is estimated that the increase of the concentration of PM_{10} to 10 μg/m^3 daily mortality from all causes would increase by 0.2–0.6% (Samoli et al. 2008). In conditions of constant exposure to $PM_{2.5}$, each increase of PM concentration to 10 μg/m^3 is associated with an increase of cardiopulmonary mortality risk to 6–13% (Iii et al. 2002; Beelen et al. 2008).

The sensitive people, suffering from lung diseases or heart conditions, as well as old people and children are particularly vulnerable. Make sure to note that the evidence of the dangerous nature of PM, formed because of combustion (both from mobile and stationary sources), are more consistent and less controversial than the data related to PM of other sources (WHO 2007). The part of $PM_{2.5}$, known as black carbon, formed as a result of incomplete combustion, attracted the attention of professionals working in the field of air quality, resulted with the accumulation of the data on the health and climate harm of black carbon. Now it is believed that the health harm occurs due to the action of many PM components associated with black carbon, for example, six organic substances such as PAHs, which are known by their carcinogenic and direct toxic effects on the cells, as well as the influence of metals and inorganic salts. The International Agency for Research on Cancer classified diesel engine exhaust gases (composed mainly of solid particles) as carcinogenic (group 1) for humans. The same list includes some PAHs and PAH substances as well as solid fuel combustion products in the household sector (Downs et al. 2007; Schindler et al. 2009; Erp et al. 2012).

Urban smog is formed mainly via two factors—the first is the increase of atmospheric pollutants, including particulate and gaseous pollutants, industrial pollution, vehicle exhaust emissions, and straw combustion that can release large amounts of pollutants, especially particulate matter, leading directly to reduced visibility and the second one is the diffusion of meteorological conditions such as horizontal and vertical wind. In recent years, the rapid development of urban construction leads to making of many kinds of high buildings, acute increase of surface friction coefficient, significant decrease of wind speed and static horizontal wind phenomenon, and significant increase of static horizontal wind phenomenon. This is not conducive to urban air pollutants diluted with peripheral expansion, but it results with the gradual accumulation of high concentrations of pollution in urban areas. The smog, covering like a lid, hangs over the city and appears over the city at low altitude, higher than the temperature inversion phenomenon, not allowing the pollutants to spread easily.

Haze has an important effect on the land, sea, and air transportation. Haze reduces visibility, leading to accidents. Studies have shown that the haze can reduce crop yields, mainly due to the fact that the haze reduces the sunlight reaching the ground so that the total solar radiation amount reduces for good and sunshine time shortens, resulting in reduced photosynthesis.

According to foreign literature, in December 2002, Chinese Academy of Meteorological Sciences translated "Gray Haze" as "haze." Later, Beijing forum named "haze formation mechanism of regional atmospheric and climate effect and predictions seminar" was held, reporting on the proposed concept of haze, deeming that human activity increases the formation of atmospheric aerosols near urban areas, thus formally proposing the "haze" concept. With the unprecedented rapid industrialization, many major Chinese cities are facing with reduced visibility problems caused by serious regional air pollution from large regional perspectives. There are four districts clear of haze. They are the Huanghuai region, the Yangtze River valley, the Sichuan Basin, and the Pearl River Delta. The research, held for half a century (in 1951–2005), shows temporal and spatial variations of haze. The number of haze days has increased significantly in the last 80 years. It has also shown that the number of haze days is usually above normal in December and January, while the number of haze days in September is the least one. Regional haze, continuing for a few days, concentrates in central Liaoning, Sichuan Basin, the North China Plain and Guanzhong Plain, and southern regions more. The regions are more affected by the storms (Wu 2012).

1.2 Existing Research Work

1.2.1 Research on Atmospheric Remote-Sensing Satellite Technology

Foreign Research on Atmospheric Remote-Sensing Satellite Technology

(AOD) Inversion

The aerosol optical thickness (AOT) is the key parameter for the evaluation of the atmospheric air pollution studies with optical column, which is also the most important unknown element of each atmospheric correction algorithm for solving the equations of transport and removal of atmospheric effects, satellite remote-sensing images (Hadjimitsis et al. 2004). A measure of the loading of the aerosol in the atmosphere is also called aerosol optical depth (AOD). Higher values indicate higher concentration of AOT aerosols and, hence, air pollution (Retalis et al. 2010). The use of Earth observation-based monitoring and determination of AOT can be directly or indirectly used as a tool for assessing and measuring the air pollution level. Measurements of PM_{10} and $PM_{2.5}$ are associated with the values of AOT as shown by Hadjimitsis et al. (2010) and Lee et al. (2011a, b).

"Aerosol optical thickness" is defined as a degree to which the aerosols prevent the transmission of light. The aerosol optical depth or optical thickness (τ), a measure of atmospheric transparency at the wavelength λ, is defined as the integral extinction coefficient (δ_λ). It is the sum of the absorption and scattering of the vertical column of unit cross-section $N(z)$ and the top of the atmosphere (H):

$$\tau_\lambda = \int_0^H \delta_\lambda N(z)\mathrm{d}z \tag{1.1}$$

So in order to retrieve the AOD, the extinct contribution from clouds, gases, and molecules have to be derived and subtracted from the atmospheric optical depth (Levy et al. 2013).

The study on remote-sensing $PM_{2.5}$ originated in applied research and retrieval of aerosol optical depth AOD. In this regard, domestic and foreign students did a lot of work. The AOD international inversion began in the 1970s. Griggs (1975), Fraser (1976), and other radiative transfer model simulation scientists showed that in the absence of cloud plane parallel atmospheric conditions, the existence of monotonic correlation between radiation and aerosol optical thickness on the top of the atmosphere is visible and infrared.

After the 1980s, based on the satellite images of the land surface atmospheric correction needs, terrestrial remote-sensing aerosol algorithm has been developed. During this period, the multichannel reflectivity algorithm (Fraser et al. 1984; Durkee et al. 1986), the structure function method (Tanré et al. 1988a, b), the thermal infrared contrast method (Legrand et al. 1988), the dark as an element method (Kaufman and Sendra 1988), and the land and sea contrast method (Kaufman and Joseph 1982) have also been obtained. Among them, the dark, as an element method, has become the most widely used one against land remote-sensing aerosol through continuous improvement.

Since the 1990s, the aerosols effect on climate and environment is increasingly valued by the international community. So, the new satellite sensors began to consider the need for remote sensing of aerosol in the design. So, the aerosol retrieval algorithm has become a new development. The Along-Track Scanning Radiometer (ATSR-2 (ESA, 1995)), Multi-angle Imaging SpectroRadiometer (MISR (USA, 1999)), multi-angle reflectance information Advanced Along-Track Scanning Radiometer (AATSR (ESA, 2002)), and other sensor applications for remote-sensing aerosol inversion produced a reflectivity angle distribution method (Martonchik and Diner 1992; Veefkind et al. 1998), which does not make only AOD inversion algorithm a certain development but also achieves a win–win situation in order to obtain the spectral distribution of the aerosol particle scale. In the late 1990s, MODIS data began to be widely used in global aerosol optical depth inversion (Kaufman and Remer 1994; Tanré et al. 1999) with the Moderate-Resolution Imaging Spectroradiometer (MODIS (USA, 1999)) emission sensor. The National Aeronautics and Space Administration (NASA) MODIS aerosol optical thickness of global business products MOD04_L2 used a dark pixel algorithm to provide part of the world's oceans and over the land with aerosol optical depth product, a spatial resolution of 10 km (nadir).

In the twenty-first century, with the deepening aerosol research, remote sensing can be used in aerosol research, and satellite sensor performance continues to improve. Moreover, the inversion algorithm is also expanded, as well as a new generation of dark blue (Deep Blue) (Hsu et al. 2004, 2013) algorithm is being developed. The algorithm of the distribution of aerosol in arid and semiarid areas is suitable for high resolution, and it has been operational.

In recent years, foreign remote-sensing aerosol research mainly have been using MODIS, MISR, and other multispectral sensors. Most of the achievements mainly include the following three aspects:

1. Using MODIS/MISR, ADEOS/POLDER, ENVISAT MERIS/AATSR, and other sensors in different areas for AOD inversion experiments, using the inversion algorithm to improve and innovate the retrieval accuracy constantly, and making the use of solar or global aerosol spectrometer network (AERONET) observational data validated.
2. The comparative research on inversion and aerosol products and the comparative research on different sensors getting data of the AOD inversion improve the inversion algorithm and thereby improve the retrieval accuracy.
3. The inverse correlation analysis results in AOD and other geographical, environmental, and climatic factors of various statistics and explores the interaction mechanism between them, trying to monitor the AOD for the environment and climate change.

After the sun aerosol backscatter-based remote monitoring, there were the achievements based on the apparent change in reflectivity due to aerosol retrieval of aerosol. A smaller table of secret AOD data obtained under the conditions of atmospheric molecular scattering can be approximated as the reflectivity R_m and R_a aerosol scattering and surface reflectivity R_S of the atmosphere contribution.

$$R(\theta) = R_m(\theta) + R_a(\theta) + R_S(\theta) \quad (1.2)$$

Here a scattering angle θ depends on the perspective and the sun geometric plane.

$R_a(\theta) + R_S$ is a function of the aerosol optical thickness $\tau_{a.}$ Aerosol single-scattering albedo ω scattering in overall extinction proportion of representatives and aerosol scattering phase function $P(\theta)$ represent the scattering angle distribution. When τ_a is smaller, there is a function:

$$R_a(\theta) \infty \tau_a \, \omega P(\theta) \tag{1.3}$$

In addition, the different wavelengths and angles of the optical information include aerosol, aerosol spectral distribution (Tanré et al. 1996; Martonchik et al. 1998), the shape of the aerosol (Kalashnikova and Kahn 2008), and the aerosol composition (Kahn et al. 2001; Kaufman et al. 2005; Sun et al. 2013). Inversion, due to atmospheric scattering, is also needed to consider the vertical distribution of aerosols. In the shortwave band, soot, dust, and organic carbon aerosols absorbing impacts can reduce or even reverse the effects of aerosol scattering. Non-absorbing aerosols aren't generally observed on the bright surface (such as snow, ice, and clouds). On the bright surface, the use of aerosol absorption band change, particularly in the UV and blue wavelengths (Herman et al. 1971), and absorbing aerosol can be clearly detected.

The main problem is troubleshooting remote-sensing aerosols affecting clouds' apparent reflectivity, as well as determining the type of surface reflectance and aerosol reflectance. Inverting the development of land-based aerosol using satellite remote-sensing data, you can simply put satellite remote-sensing method to retrieve aerosol into the following categories.

Dark target. The principle is as follows: first, in addition to dust, the aerosol optical thickness decreases with increasing wavelength, so it is small 3–30 times in the shortwave infrared (2–4 μm) aerosol optical thickness ratio corresponding to the visible light (0.47 and 0.66 μm); second, it depends on the sun's surface reflectance spectral band wavelength. The basis of this principle is that the reflectivity of a large number of dense vegetation in the red and blue bands is relatively low (0.01 to 0.02), such as tropical forests and deciduous forests. It's likely that most of the precise types of information are obtained from the sensor surface and in the atmosphere contribution correspond separately. Thus, this method can be used with more chunks of dense forest cover or vegetation to get information about aerosols. The method is called dark target or dark pixel method. It also can be called dense vegetation law, mainly for low spatial resolution sensors (such as AVHRR, MODIS, etc.). This method generally yields ideal results on dark surfaces (Matsui et al. 2004; Remer et al. 2005; Levy et al. 2010a, b; Sayer et al. 2015a, b), but it may also lead to larger deviations, especially in coastal and arid regions (Abdou et al. 2005; Die et al. 2013).

Structure function method. On a light surface, the two-sided nature aerosols for satellite-measured reflectance effects can both make a significant increase and a significant reduce, ranging from high surface reflectance and aerosol, monitoring the impact of interference the resulting limitations. In this regard, some scholars have proposed a new aerosol monitoring method—structure function method (Tanré et al. 1988a, b; Holben et al. 1992; He et al. 2012). The structure function method is used for a group of images, which requires a relatively clear image, i.e, to obtain thick and reliable aerosol optical thickness, and use relevant methods to estimate the aerosol optical thickness of this image in advance (or by observing and interpolating the ground to obtain its optical thickness), and then use this as the basis to calculate the aerosol optical thickness of other images in the group. Advantage of the structure function method is limited by the size of the surface reflectance less. The structure function method is usually applied to arid and semiarid areas.

Polarization algorithms. Atmospheric aerosols and atmospheric molecules interacting with the incoming solar radiation, in addition to scattering and absorption of incident radiation, can also make the polarization of the incident radiation, satellite information to obtain aerosol scattering polarization characteristics by measuring the atmosphere, the inversion aerosol optical thickness, the spectral distribution, the refractive index, the phase function, and other parameters.

UV algorithms. Ultraviolet observations for using a surface signal are minimized (Herman et al. 1997; Torres et al. 2013). The UV aerosol absorption method is more sensitive, though it inverts tropospheric aerosols less sensitively than the inversion of the boundary layer aerosols. However, they have the same inversion effect over both land and water surfaces (Torres et al. 2007a, b).

Multi-angle algorithms. The suppositions that the angle variation of surface radiance is not wavelength-dependent are based on different angles of surface radiance changes. They can be separated from the total contribution of surface radiance from the top of the atmosphere (Martonchik 1997; Diner et al. 2005; Dubovik et al. 2011).

Dark blue algorithms. Hsu et al. (2006, 2013) leans on the red and blue aerosol optical thickness bands, on the zenith radiance considerable contribution, and on the blue surface reflectance library (Deep Blue) algorithm. He uses the established images of surface reflectance from the SeaWiFS rate library. The AOD is used only in a small blue data inversion; large red and blue light is the integrated use of data inversion. This method is successfully applied in arid and semiarid Sahara, the Arabian Peninsula, and the like.

Particulate Pollution Remote Sensing

In recent years, scholars from various countries lean on satellite remote sensing to estimate the concentration of particulate pollutants near the ground and to monitor the extensive research conducted on regional pollution; the main ideas are either directly or indirectly associated with the AOT model particle concentration near the ground by means of the establishment of other parameters. According to the difference correlation model conceived by atmospheric factors, these studies can be divided into the following categories.

1. AOT and particulate matter concentrations near the ground are directly related to the model

   In order to explore the AOT estimate particle concentration near the surface of feasibility, some scholars' terrestrial observations of the AOT and $PM_{2.5}$, and PM_{10} are concentrated near the ground for a preliminary correlation analysis. For example, Chu et al. (2003) compared the observations of northern Italy AERONET AOT PM_{10} concentrations near the ground. The correlation coefficient *R* was 0.82. Slater et al. (2004) used a self-photometer inversion of AOT and compared it with the ground $PM_{2.5}$. The correlation coefficient *R* was 0.87.

The studies demonstrate the AOT estimate particle concentration near the ground feasibility.

Other scholars compared directly the AOT and particulate matter concentrations near the ground satellite remote sensing and got good results. For example, Wang and Christopher (2003) from Alabama compared the standard of MODIS AOT ground and $PM_{2.5}$ concentrations. It was found that the correlation coefficient *R* between the MODIS AOT and the corresponding $PM_{2.5}$ hour average was 0.7, and the correlation coefficient *R* between the monthly averages was 0.92, which not only revealed the great potential of MODIS air pollution monitoring but also showed that there was a greater consistency between satellite remote sensing AOT and near-surface particulate concentrations on larger time scales. Kassteele et al. (2006) compared MODIS AOT standard of PM_{10} and ground-based observations across Europe, where some correlation ($R = 0.58$) was. Leaning on the correlation between the two, they used the MODIS AOT estimated distribution of $PM_{2.5}$ in the country and ground observation interpolation results to compare them to the large spatial scales, to analyze the applicability of satellite remote sensing, and to estimate near-surface particulate matter.

2. Consider the vertical distribution of aerosols and the relative humidity of the relevant model.

   When estimating near-surface area of particulate matter in Europe, Kassteele et al. (2006) leaned on the assumption of negative exponential aerosol vertical distribution. The use of the European Centre for Medium-Range Weather Forecasts (ECMWF) released atmospheric boundary layer height and relative humidity data. The vertical correction and humidity correction of MODIS AOT greatly enhanced the correlation between AOT and $PM_{2.5}$ and PM_{10}.

   Wang et al. (2010) assumed that the vertical distribution of aerosols was a negative exponential decay. Firstly, the correlation model of AOT and near-surface particle concentration was given, and the validity of the vertical correction and humidity correction methods were respectively verified by the measured data; then, the atmospheric boundary layer height was detected by laser radar and the measured relative humidity were used to invert the foundation; AOT performed vertical correction and humidity correction, and the corrected R^2 for AOT and PM_{10} and $PM_{2.5}$ reached 0.65 and 0.62, respectively. Finally, leaning on this, it estimates a foundation of data given by the satellite remote sensing of AOT estimate particle concentration near the ground empirical model. Using MODIS 1 km AOT estimated near-surface concentrations of PM_{10} and $PM_{2.5}$ in Beijing, as well as observed and compared values, we can say that both of them have a high correlation level ($R^2 = 0.47$) and higher consistency in larger spatial scales and longer time scales.

   Engel-Cox et al. (2006) and Hutchison et al. (2008) measured correlation with a laser radar ground. The airborne aerosol vertical distribution data, significantly improving the MODIS AOT and $PM_{2.5}$ correlation results near the ground level, showed that the vertical distribution of aerosol-based AOT for vertical revision had obvious advantages.

Liu et al. (2004a, b) pioneered the use of atmospheric chemical transport models simulating the entire floor at each grid AOT correlation between the concentration of particles near the ground, and basing on the use of satellite remote sensing of aerosol optical thickness in the United States (mainland part), annual average $PM_{2.5}$ concentrations were estimated, and good results were achieved. After that, Liu et al. (2007) and Lee et al. (2011a, b) leaned on atmospheric chemistry model to simulate aerosol vertical profiles of aerosol optical thickness of satellite remote-sensing vertical height revised; it resulted in the extinction contribution within the air layer near the ground aerosol surface; the results showed that the method could reflect the recent changes in the concentration and analog ground particles better. Choi et al. (2009) used a similar method in the annual average and seasonal average scale effects of the East Asian region MODIS AOT and near-surface PM_{10} and $PM_{2.5}$ correlation.

3. Consider a variety of factors related to meteorological model.

   Gupta et al. (2006) and Christopher and Gupta (2010) analyzed the sensitivity of AOT and ground particle concentration and correlation between cloud cover, wind speed, mixing layer height, relative humidity, and other weather factors and noted that, preliminarily, satellite remote sensing near ground concentration of particulate algorithm should have regard to environmental and meteorological factors.

   Liu et al. (2004a, b, 2007) used a common linear regression model (general linear regression model, GLM), the AOT, boundary layer height of satellite remote sensing, and the relative humidity as major factors, as well as a number of geographical and environmental factors as cofactors, including geographical, season, land surface (rural/suburban/urban) properties and the distance from the coastline, and the like. Through a large number of samples, regression analysis of the data was obtained for each factor of the regression coefficients of $PM_{2.5}$ concentrations near the ground. Based on this model, the MISR AOT data of $PM_{2.5}$ ground near the Eastern United States was estimated and compared with the observed values. On this basis, as an example of the abovementioned factors of the GLM model for a certain degree of adjustment, Liu et al. (2007) considered that the St. Louis area would be the main AOT factor; boundary layer height and wind speed and air temperature would be replaced with such cofactors as wind direction and season, as well as the use of model-adjusted comparison of MODIS and MISR AOT on the ground near the ability to estimate $PM_{2.5}$.

Home Development Status of Atmospheric Remote-Sensing Satellite Technology

1. Aerosol Optical Depth (AOD) Inversion

   In China, the remote sensing of aerosol retrieval work began in the mid-1980s, and it obtained a lot of research results. Basing on the extinction angle-scattering distribution, Qiu (1995) proposed an aerosol remote-sensing spectral inversion principle for experimental studies, and he provided a full-band direct sun radiation information to determine 0.7 μm wavelength aerosol optical thickness. Gu et al. (2001) used the ADEOS/POLDER polarization data and studied remote-sensing Inner Mongolia grassland where there was tropospheric aerosol. Zhang et al. (2003) used GMS geostationary meteorological satellite data to retrieve aerosol optical thickness over the lake, and more than five lakes nearby photometer band remote sensing to compare the results of the inversion. The results are verified using this last inversion method in Mainland China over the lake 25 aerosol optical thickness. The results showed that the satellite retrieval of aerosol optical thickness of lake AOD inversion over the continent to provide information was a feasible method of study of aerosol optical depth over the continent.

In the twenty-first century, after the successive launch of Terra and Aqua, the MODIS spectral data for its wide range of data reception is simple and has a high update frequency and free use of the characteristics. Due to the work of domestic and foreign scholars, aerosol study provides more extensive information. In 2001, Li et al. (2003) studied the use of the MODIS data on Beijing and Hong Kong urban aerosol inversion algorithm; he also studied the use of NASA's aerosol products, obtained seasonal distribution characteristics in East China, and combined it with a synoptic method for air pollution cases. An important example is the inversion of urban land aerosol. Wang et al. (2015a, b) used the 6S mode MODIS blue, red, and infrared channels of surface reflectance of the planet. The results of the aerosol optical thickness sensitivity test show that in Lanzhou and the surrounding areas, the correlation between the blue channel surface reflectance and the infrared channel surface reflectance is more in line Kaufman.

Many domestic scholars studied remote sensing of aerosols. Mao et al. (2002) studied remote sensing of aerosols in the Beijing area; the use of NASA MODIS land aerosol releases ready data products. The data corresponding to the measured data obtained with the ground photometer for comparison were analyzed in time and space. The aerosol optical thickness points were described in Beijing. Li et al. (2003) compared the sun-photometer measurement on aerosol optical thickness in Beijing; NASA released MODIS aerosol products; the reliability of the satellite remote-sensing products was verified; and the air pollution index will be inferred from the daily average concentration of particulate matter of theory and the MODIS data of aerosol optical thickness product correlation analysis. Li (2003) used the MODIS data on the Beijing aerosol dark-like element method and structure function method. He made two comparative studies of different kinds of inversion method. Wang et al. (2003a, b, 2008) leaned on the MODIS aerosol inversion test in Jiangxi and Beijing areas, using the dark target algorithm. Sun et al. (2006), Wang et al. (2009a, b), Zheng et al. (2011, 2016) leaned on the use of domestic environment of HJ-1 satellite and aerosol inversion, respectively.

2. Remote-Sensing Particles of Pollutants

Currently, particulate matter by satellite remote-sensing research conducted by scholars is still relatively small. Through improved MODIS inversion algorithm, Li et al. (2005) made the MODIS AOT spatial resolution of 1 km and applied it to monitor atmospheric pollution in Hong Kong. He found that 1 km AOT cannot only reflect the complex changes in the pollution degree in urban areas. He compared a 1 km AOT to a 10 km AOT and its higher correlation to PM_{10} level. It indicated that the high-resolution satellite remote-sensing data had a huge potential for urban pollution monitoring. Guo et al. (2009) compared the concentrations of a 10 km AOT in East China showed by MODIS with surface particulate matter. He found that both of them were significantly correlated, and their near ground particles' hourly average ratio of daily average and MODIS AOT related was higher.

1.2.2 Experience of Studying Urban Areas Using Thermal Images

Modern cities occupy only 2% of the land, but they are already home to more than half the world's population, consuming ¾ of the world's natural resources. The cities, where population and production are concentrated, are the centers of powerful impact on the environment. Water pollution, air pollution, and soil pollution are primarily associated with cities. Problem of the ecological state of urban landscapes is essential. It is a complex, taking into account the pollution of water, air, soil environment, and economic aspects of urban placing facilities and the problems associated with the level of comfort for people's life.

Thermal pollution is often caused by unsustainable urban space layout. The thermal anomalies caused by urban agglomerations are a common phenomenon of the modern world. In cities, such sources of anthropogenic heat as industry, transport, and objects related to the housing and communal services are concentrated. Most of urban objects consist of materials, which actively absorb solar radiation (asphalt, concrete, granite, and similar thick materials), leading to heat accumulation within the city and its active radiation in the environment: urban buildings and structures become sources of intense heat radiation. A lot of thermal anomalies, related to urban areas, form the urban heat island. The urban heat island (UHI) is a phenomenon associated with an increase of air temperature in the city comparing to its surrounding areas.

The urban environment differs the surface temperature from the temperature of the natural environment, which causes the feasibility of attracting materials of heat shooting during ecological studies of urban areas.

Formation of the "heat island" is also associated with features of the Earth surface geometry in the city. Tall buildings have a large surface area to reflect and absorb solar radiation, which increases the intensity of heating of urban areas. This phenomenon is called the "urban canyon effect." Another feature of the contribution of the buildings to the formation of "heat island" is that winds are blocked, causing to a decrease in the intensity of convective cooling. Cars, industry, and other sources also contribute to the formation of excess heat. The high level of pollution in urban areas may also lead to the effect of "heat island," as many types of pollutants change the radiative properties of the atmosphere. Heat losses in the energy sector are the second leading factor. Large areas are changed when urban centers are growing. The average surface temperature increases within the borders.

Heat island city is characterized by a pronounced daily dynamics: the largest values of the temperature difference between the city and the suburbs in the evening and at night fill the heat accumulated during the day. When it comes to seasonal dynamics, it should be noted that the heat island appears both in summer and in winter. There are heat island related to the air temperature and heat island related to the Earth' surface temperature (surface of heat island) and to the lack of vegetation in urban areas.

For example, the next aspects are studied: the daily dynamics of the thermal field of the city, the influence of use of the city land on spatial-temporal dynamics of the local thermal anomalies of the city, and the correlation changes in air temperature and surface temperature within the city area. The simulations of urban heat island as surface temperature and air temperature, modeling the energy balance of the city, and a comparison of temperatures within and outside the city are carried out. The differences of the intensity of thermal radiation of objects in the day and night time are studied. When comparing the calculations according to the visible and near-infrared ranges of the values of vegetation index (normalized difference vegetation index (NDVI)), the correlation between the abundance of vegetation and the intensity of the thermal radiation characteristics of land use is also studied. In many works, the main source of information is the data of satellite images in the thermal infrared range.

The data of thermal infrared survey is used in geographical studies of different spatial coverage of urban heat islands; the system determines the range of characteristics: the width of visibility and the spatial resolution of images. For example, multichannel radiometer AVHRR NOAA satellites and the MODIS, set on the Terra and Aqua satellites with a review of the visibility near 2000–3000 km, provide images of low spatial resolution (about 1 km), which are used in studies of large spatial coverage (Zhu and Blumberg 2002; Weng et al. 2004; Lu and Weng 2006; Carrer et al. 2014; Kotarba 2016). This allows to estimate the total power and the length of the heat island (Voogt and Oke 2003; Weng et al. 2006; Zhou et al. 2008; Qiao et al. 2013) and its impact on surrounding areas, as well as to compare the size and intensity of the heat islands in different cities. The advantage of such materials is in the high repeatability of shooting; a large number of the set of channels, however, the spatial resolution of 1 km, are not enough to study the spatial homogeneities in the heat island.

Another approach is the use of heat island study images with a spatial resolution of not less than 100–120 m, such as ETM+/Landsat-7 and ASTER/Terra. The Landsat satellite images are some of the most common materials of Earth remote sensing in geographical research at a regional level. In particular, it applies to the images in the thermal infrared range (Yang et al. 2004; Srivastava et al. 2009; Reuter et al. 2015). These images allow us to see the internal spatial structure of urban heat islands, controlled over time, and space of the development of local thermal anomalies, as well as to evaluate the thermal effect of various urban facilities on each other. The Landsat satellite from urban heat island studies pictures used as supplementary material (Weng 2009), as well as the core.

There are examples of multi-data high and low resolution thermal range (MODIS, ASTER, Landsat), obtained from different satellites, in studying the features of heat islands (Weng and Quattrochi 2006a, b; Liu and Zhang 2011; Lazzarini et al. 2013). Combined use of different spatial resolution data is very valuable for what makes it possible to take advantage of the important coverage and high-resolution images. Usually, there are recommendations on the use of the images with different resolutions to study heat islands at different spatial scales among the results of such work.

Different types of classifications of satellite images in the thermal IR range (Lu et al. 2015) are used to study the spatial characteristics of a heat island. Typically, such methods are used to study "land use" and to create "maps of Earth covers" (land cover/land use). Involved pictures in different spectral ranges are used more often to research land cover/land use; one of the pictures illustrates heat. Maps of land cover/land use are created in analyzing correlation between the types of "land use" and thermal anomalies (Zheng et al. 2016). It denotes the existence of the correlation between the structure of heat islands and land cover (land cover/land use), and the possibility of using thermal images as a reliable source of information on the nature of land is used again (Wan and Li 1997).

Different options of modeling the thermal characteristics of the surface and extracting heat radiation parameters occupy a significant part of the scientific research papers of spatial characteristics of heat islands. One of the most frequently raised problems is extraction of the land surface temperatures (LST) of the Earth surface temperature from thermal images (Sobrino et al. 2004; Ding and Hanqiu 2006; Zhao et al. 2009; Oguz 2013; Wang et al. 2015a, b). The LST values can be obtained from the device that shoots simultaneously in several thermal channels (the algorithm, "splitting windows transparency," the split-window method, and SWF). Difficulties arise if the photographing is performed in the one channel, as it is performed in the ETM+ and TM systems. The problem is tried to get around, but it is not always possible; there are many options for extracting algorithms LST images of ETM+, but they give only approximate results. Different methods are used to get real results of the algorithm as close as possible. For example, the accuracy of the algorithm LST extraction may be estimated by comparing it via ground observation; option with the smallest difference is selected as a working version. But the ground observation should be carefully planned; otherwise, their objectivity as a means of validation of algorithms can be called into question. For example, when using the algorithms, as a ground-based measurement data of meteorological parameters in weather stations, the microclimatic situation—the weather station is often located in the city parks—local "islands of coolness" of the city should be taken into account.

Great attention is paid to the studying the correlation between vegetation capacity and the intensity of the thermal radiation. When comparing two such images, a pronounced feedback is always found. The correlation is often evaluated by the LST and the NDVI image analysis (Sun and Kafatos 2007; Julien and Sobrino 2009; Song et al. 2014). Using the NDVI and the LST provides additional information on the Earth surface, such as evapotranspiration and soil moisture (especially in areas with sparse vegetation). The results of the joint analysis of LST and NDVI images are often used to assess the impact of urbanization on environment; sharing the LST and NDVI maps and land cover/land use gives the most complete picture. Often, however, it was pointed to insufficient efficacy NDVI as a means of assessing the capacity of vegetation cover in the city. Other methods of assessing the impact of urbanization were used. For example, a joint analysis of the LST and vegetation fraction from the model of the spectral decomposition was carried out (Tan et al. 2012).

The researchers engaged in both seasonal and diurnal spatial-temporal dynamics of the heat islands (Vinnikov et al. 2008; Xu et al. 2010; Shi et al. 2013; Wang et al. 2017). Sensor data, such as AVHRR/NOAA, TM/Landsat, and MODIS/Terra as well as aircraft and ground vehicles, is used. The study of daily dynamics allows to evaluate the thermal characteristics of different urban facilities and changes of the amplitude of daily temperature, as well as to identify the objects that form thermal anomalies at different times of a day and to determine the time of a day, which is more suitable for the heat shooting for a particular purpose (Peterson 2003). However, when using thermal images from Landsat, it is hard to illustrate with the help of satellite images of daily change of the heat island due to features of satellite orbit (orbit is sun-synchronous orbit; when connecting with it, the satellite passes over a certain place in the same time of the day). Night Landsat images are available in the archives, but they are few. The study of seasonal dynamics allows to reveal structure changes of the heat island during a year, features of the dynamics of local thermal anomalies, and the thermal characteristics of anthropogenic and natural objects on the scale of the whole year.

There are comprehensive studies combining constructing the LST images, the studying of seasonal and daily dynamics in urban areas under different climatic conditions, the studying of statistical analysis, the communication studying of the LST images, the NDVI, and maps of land cover/land use, and the studying of heat fluxes (Li et al. 2008). Both pictures of low spatial resolution and of a large cover, as well as images from Landsat in such works, were used. This type of studies provides a multilateral assessment of the heat island phenomenon and makes a significant contribution to the city's climate research and the impact of urbanization to climate research.

1.2.3 Research Status of Atmospheric Particulate Matter Pollution Levels

Because of dangerous effect of atmospheric particulate matter on cities and villages, especially in urban respirable particulate matter pollution, now atmospheric particulate matter has become a primary pollutant in many cities. It also has a lot of adverse effects on the environment. Since the 1990s, in the United States, Europe, and Japan, all the atmospheric particulate matter, including respirable particulate matter, is a subject of an extensive and in-depth study; they are formed from the source of generation to enter the air diffusion; then they enter the body and many other areas; we have made many important discoveries and conclusions. In April 2008, the council of the European Parliament adopted the "Directive on ambient air quality and cleaner air for Europe," and revised air quality regulations for the first time to set the standard $PM_{2.5}$ limits by 2015 when $PM_{2.5}$ concentrations in the air should not exceed 25 μg/m^3. Before June, 2010, the regulations had to become a national decree member state of the European Union. In 1997, the World Health Organization (WHO) published a new "Air Quality Guidelines" (AQG) and air quality guidelines for a particulate matter (PM); as for sulfur dioxide, nitrogen dioxide, and ozone, the standard annual average concentration was 10 μg/m^3, 24-h average concentration was 25 μg/m^3 in 2005 (2005 global update of a risk assessment summary). In 2003, in Australia, $PM_{2.5}$ was admitted to the ambient air quality standards; its daily average concentration limits and annual average concentration limits were 25 and 8 μg/m^3; almost all developed countries increased $PM_{2.5}$ environmental quality standards.

Developed Western countries and regions have carried out a research work on the ambient air quality standards revised system, with reference to the Chinese ambient air quality standards and amendments. One of the researches on atmospheric particulate matter started in the late 1970s established air quality standards for the TSP, and the "ambient air quality standards" promulgated in 19% (GB3095-1996) and modified and added PM_{10}. For example, in October 1999, in Beijing, the PM_{10} data had been reported; other major cities had been reported the PM_{10} data in 2000 or 2001. At the time, China had not got $PM_{2.5}$ ambient air quality standards; it was the relative lack of research on $PM_{2.5}$ observational approaches; the data system was also poor. In 2012, after the adoption of the GB3095-2012, the hourly and daily monitoring of air quality began in all major Chinese cities on the basis of the data of the six main pollutants of air (PM_{10}, $PM_{2.5}$, SO_2, NO_2, CO, and O_3)**.** China is one of the countries in the world where serious air pollution and the city's air pollution are more serious. The average concentration of particles and dust in the atmosphere is greater than the national standard level two; it exceeds the upper limit value of the World Health Organization equal to 90 μg/m^3 from 1 to 7 times (Wang 2005).

In terms of atmospheric particulate matter, competent scholars have made a lot of research, showing that particulate pollution of the city is quite serious. According to urban air quality daily data, Tian et al. (2005) used the SEPA and published the study of the spatial and temporal distribution of air pollution in major cities in 2001–2004. The study results showed that the primary pollutant in major Chinese cities was PM_{10}; in northern parts of southern cities, spatial and temporal distribution of pollutants was obvious; urban air pollution in winter, spring, summer, and autumn was heavy. At the same time, it shows that the effects of sand- and dust storms on China's urban air quality, in cities in northern China, especially in northwest cities, affected by the dust weather, is relatively large, and the PM_{10} pollution is relatively serious.

The scholars have made extensive research on the PM_{10} level in Chinese cities; many studies have shown that there is seasonal difference of city particle concentration in China, but the local meteorological conditions, energy structure, different functional areas, and geographical location have been also closely linked to influence factors. Scholars' researches such as Ta et al.'s (2004), Chu et al.'s (2008), Yu et al.'s (2010), and Filonchyk et al.'s (2017) showed that the PM_{10} mass concentration distribution in Lanzhou had obvious seasonal variations, which were generally the highest in winter, lower in spring, lower in summer and autumn, than in spring, and lowest in summer. The main reason was heating in winter and spring and north frequent sandstorms in spring. Lanzhou is a typical northern Chinese polluted city, where the winter heating period is when the number of coal-fired boilers surge, coupled with less rainfall, and prone to temperature inversion phenomenon, resulting in a more stable atmosphere and pollutants easy to spread, coupled with frequent spring north sandstorm; so, in the winter season, Lanzhou levels of atmospheric particulate matter are relatively high. An unstable layer of atmosphere, reducing emissions of pollutants in summer and rising the level of sun radiation and warming the ground quickly, is often formed near the ground to promote vertical air convection. The increasing in summer rainfall for atmospheric particulate matter also played a role in erosion.

Particle observations of scholars of many cities showed that its fine particle pollution situation was quite serious; in 1995 and 1996, Wei et al. (2001) monitored Guangzhou, Wuhan, Lanzhou, and Chongqing, where the annual average concentration of $PM_{2.5}$ was 57–160 μg/m^3. From 2005 to 2006, Jun et al. (2009) suggested to observe atmospheric particulate matter in Beijing: PM_{10} concentrations were highest in spring and autumn, and they were low in winter and summer, reflecting the Beijing PM_{10} pollution and meteorological conditions in two seasonal variations under the role. The $PM_{2.5}$ concentration was highest in spring, autumn, and winter and minimum in summer. In 2006, at two air quality monitoring stations in Hangzhou, Zhen (2010) received data on the mass concentration $PM_{2.5}$ and $PM_{10,}$ showing that the annual average concentration exceeded the National Ambient Air Quality Standard (NAAQS) with those of 77.5 and 111 μg/m^3, respectively. Filonchyk et al. (2016) analyzed the pollution in the cities of Gansu Province in the summer period of 2015, and the results showed that the ratio of $PM_{2.5}/PM_{10}$ was 0.405. These studies indicated that a fine particle became a major component of PM_{10}.

The state of atmospheric pollution in China is complex; rapid urbanization and economic growth have caused a combination of primary pollution (such as SO_2 and NO_x pollution) and secondary pollution within a short period of time, showing the characteristics of complex atmospheric pollution. SO_2 pollution problems and vehicle exhaust pollution problems have not been resolved one by one; atmospheric particulate matter pollution is very serious. Atmospheric fine particles and the large contribution of secondary pollution, as well as of spatial and temporal changes, could easily result in regional pollution and source apportionment; pollution control has resulted in great difficulties. Therefore, while the issue is still not resolved, the coarse particles and the fine particles are studied; it is bound to face serious difficulties in research foundation plaque and needy.

The research of particle size distribution of the particulate matter was the earliest and the most basic research of the particulate matter. The particle size distribution refers to the proportion of particles with different particle sizes in a certain particle group; in fact, all the physical and chemical properties of atmospheric particles are related to their particle size distribution, therefore, the research on the particle size distribution of inhalable particles has always been one of the most fundamental and most important research topics in the field of respirable particles. Results of the study showed that different regions had different particle size distribution in different seasons and at different times, while its $PM_{2.5}$ in PM_{10} proportion was also different.

Particle size and its source and the formation of atmospheric particulate matter have a close relationship. According to modern physics model of atmospheric particles, atmospheric particles divided into three particles can be expressed as modal results. Particle size of less than 0.05 μm particles is called as a love root nuclear membrane, the secondary particles of the primary particles and gas molecules and the main source of combustion generated by a chemical reaction. Over time, nuclear membrane particles of small particle size, quantity, and large surface area were easy to collide with each other by the small particles into large particles that were "aging"; it was not easy to find them in the real atmosphere. Particle size greater than 0.05 μm and less than 2.5 μm is referred to as an accumulation mode particle, mainly from the nuclear membrane.

Physical and chemical properties of respirable particulate matter are closely related to the size of particles; particle size determines the diameter particle deposition body parts. The research have shown that 10 μm of these particles can enter the nasal cavity, 7 μm or less can enter the throat, and less than 2.5 μm can be in depth and alveolar deposition and then into the blood circulation (Corn et al. 1973; Van der Zee et al. 1999; Perez-Padilla et al. 2010; Roy et al. 2016). At the same time, simple structure due to the complex structure of aggregate finer particles consists of relatively large particles of large surface area, and thus it is more easily adsorbed and harmful to human health because of some heavy metals and organic compounds, thus more toxic (Khan et al. 2008; Leung et al. 2008; Yi et al. 2011; Abuduwailil et al. 2015). The finer the particle, the longer it stays in the air, the greater the chance of being inhaled; the finer the particles, the greater the surface area, the greater the activity in the human body, the stronger the effect on lung fibrosis (dust particles having a rough edge in the lungs can irritate lung tissue, resulting in the formation of scar tissue, causing fibrosis); most of fine particles are formed by a man-made automobile exhaust and high-temperature combustion.

There can up to 107 - 108 particles per cubic centimeter in the atmosphere, ranging from a few nanometers to about 100 μm, and their size can range over four orders of magnitude. Burning sources, such as motor vehicles, power plants, and wood burning particles, are usually smaller than 1 μm in particle size and can be as small as a few nanometers in size, and dust emitted by the wind, pollen, plant debris, and the ocean discharge into the atmosphere, sea salt particles generally have a particle size greater than 1 μm. Most of photochemical reactions of secondary particles (sulfates, nitrates, and the second organic salt) are present in the particle size of less than 1 μm. The particle size of atmospheric particles not only affects its retention period, physical and chemical properties in the atmosphere, but also affects its environmental effects, or is an important basis for judging its source. Therefore, the particle size distribution of atmospheric particles is a very important part of the study. Atmospheric particulate industrial process, power plants, motor vehicle emissions and emissions from natural sources and a long-distance transmission of primary particles, and gases, which are a solid transformation reaction mixture produced secondary particles, are a very complex system.

1.2.4 Investigation of Human Health Effects of Atmospheric Pollutants

In the health effects of atmospheric particulate matter, the main revelations from a large number of epidemiological studies were made since the 1960s to the late 1980s. These studies revealed a long-term or short-term exposure of particulate matter relations to different environments; for example, a health consultation rate indicates the number of respiratory morbidity, mortality, and reduced lung capacity in between. Generally it is believed that there was a close correlation between the atmospheric particulate matter and people suffering from respiratory disease exacerbations. Epidemiological studies in Europe and the United States show the results. Hospital incidence of asthma, as well as the number of hospital deaths, would change with the atmosphere because of the PM_{10} concentration increasing.

A number of epidemiological studies associations between air pollution, especially particulate matter (PM), have been observed, and the adverse human health effects have been well documented (Fairley 1990; Schwartz 1993; Dockery and Pope 1994; Pope et al. 1999; Lee et al. 2000; Files et al. 2005; Kan et al. 2008; Ayala et al. 2012; Kim et al. 2015). PM is accompanied with a decline in external respiration function, aggravation of existing respiratory disease, cardiovascular disease, altered protective mechanisms, and even premature death. The most sensitive population group included persons already suffering from respiratory or cardiovascular disorders, asthmatics, children, and the elderly (Abelsohn and Stieb 2011).

A study conducted by Pope et al. (1995) demonstrated the correlation between the PM_{10} air pollution and cardiopulmonary mortality and lung cancer mortality. The correlation was stronger for $PM_{2.5}$ than for PM_{10}. The $PM_{2.5}$ air pollution was associated with a 36% increase in death from lung cancer and 26% increase in cardiopulmonary deaths; the risk was higher for people who were older than 65. The correlation between hospital admissions for respiratory diseases and air pollution has been assessed in the developed countries in the United States and in Europe (Atkinson et al. 2001; Wilson et al. 2005; Medinaramón et al. 2006; Giovannini et al. 2010; Tsangari et al. 2016). So, Medinaramón et al. investigated the impact of PM_{10} assessment on respiratory hospital in 36 cities in the USA; the increase of PM_{10} by 10 $\mu g/m^3$ in the warm season showed increase of chronic obstructive pulmonary disease by 1.47% and in pneumonia by 0.84%. The APHEA 2 project studied the short-term health effects of particulate matter in eight European cities. When the PM_{10} increased by 10 μg/m^3, admissions of patients older than 65 years with chronic obstructive pulmonary disease plus asthma and total respiratory diseases increased by 1.0% and 0.9%, respectively. In a Harvard University study, 8000 consecutive adults of 16 years of follow-up showed that $PM_{2.5}$ and increased mortality are higher than the correlation between PM_{10} (Darlington et al. 1997).

The positive association between ambient air pollutants and respiratory diseases was also registered in Asia (Wong et al. 1999; Li et al. 2009; Tsai et al. 2013; Tao et al. 2014). So, according to the research of Wong et al. (1999) in Hong Kong, NO_2, SO_2, and PM_{10} increases of 10 μg/m^3 were associated with total respiratory disease hospitalization and were increased by 2.0%, 1.3%, and 1.6%, respectively; a chronic obstructive pulmonary disease hospitalization was increased by 2.9%, 2.3%, and 1.9%, respectively. In Shenyang City, Liaoning Province, Wang et al. (2003a, b) showed that coal-fired air pollution could lead to cardiovascular elderly disease mortality. Under basal conditions of exposure, when the total suspended particles for every 50 μg/m^3, the total population increased cardiovascular mortality by 1.22%, while the older group of cardiovascular mortality increased by 4.3%. When studying Taiyuan particulate air pollution and mortality, Zhang and Jin-Fen (2008) showed that hospitalization rates suggested that the correlation between particulate air pollution and cancer is close; the outpatient and inpatient mortality because of human health effects of atmospheric pollution is in ascending order of severity. Zhang et al. (2014) investigated the impact of air pollution on human health in the city of Guangzhou and showed that haze and air pollution were more radical impact on people with age from 19 to 64 years old and on women. Summing it up, it can be said that each example demonstrates that haze pollution is associated with total and cardiovascular illnesses. NO_2 was the only pollutant with the highest risk of hospitalization.

1.3 Research Objectives

To achieve this goal, the following problems are investigated:

1. Making analysis of the main concentrations of pollutants ($PM_{2.5}$, PM_{10}, NO_2, SO_2, CO, O_3) in atmosphere over Lanzhou City by constructing a complete time series data on the basis of long-term data of air quality monitoring stations and using mathematical statistics, correlation, and hourly dynamics of the main pollutant concentrations to predict long-term dynamics in the main districts of the city and in the whole city; making analysis of the annual, seasonal, daily, and hourly levels of variability of pollutants, by studying the main pollutants.
2. Making analysis determining the air quality of the overall pollution level in the city based on the API (air pollution index) and the AQI (air quality index); determining the degree of air quality, the long-term and monthly dynamics of the main types of pollutants.
3. Making analysis of the meteorological conditions of air pollution, determining the rate of the natural ability of the city atmosphere to self-cleaning. Determining

the correlation between the meteorological parameters and air pollution, as well as determining the correlation between climatic factors and air pollutants.
4. Determining the overall environmental situation in the city and in the surrounding territories, on the basis of remote-sensing satellites MODIS (Terra) and Landsat 8, performing calculations of atmospheric AOD, and determining the UHI, the LST, and NDVI. Received and preprocessed satellite data would not provide just the information of atmospheric pollution in the city's annual, seasonal, and monthly progress but also would assess the thermal structure of Lanzhou City.
5. Basing on back-trajectory models of Hybrid Single Particle Lagrangian Integrated Trajectory model (HYSPLIT), Navy Aerosol Analysis and Prediction System (NAAPS), and the Ozone Monitoring Instrument (OMI), identifying the main sources of incoming of air pollutants from the dusty air masses over the city and surrounding areas, with the possibility of predicting aerosol models, and making vertical analysis of the atmosphere during dusty weather using Cloud-Aerosol Lidar and Infrared Pathfinder Satellite Observations (CALIPSO).

1.4 Significance of the Study

1. Air pollution and the necessity for studying pollution in various ways for further prediction. Due to the high saturation of the pollution sources in the cities, their level of air pollution is usually significantly higher than the suburb level of air pollution and the countryside level of air pollution. Sometimes, the adverse emission dispersion and the concentration of harmful substances may increase sharply, relative to the average value or the background of the city. The frequency and duration of the periods of increasing of air pollution will depend upon the mode of harmful emissions, as well as upon the nature and duration of the weather states increasing the impurity concentration in the surface air layer. Therefore, the study of air pollution with both primary and secondary pollutants is very important, as well as the development of different ways of investigating sources of pollution and predicting individual substances entering paths in the city limits.
2. Analysis of satellite remote-sensing systems and ground stations. Studies of air pollution include the data sources providing statistical data on the contamination level of different pollutants. It is important to compare the data obtained from land-based sources and to use the data obtained from remote sensing, as well as to compare the difficulty in obtaining data. Further results may point to the usefulness of a process of the preparation of statistical data.
3. The results obtained in the work are important for the national economy. Assessment of the conservation measures effectiveness, assessment of environmental pollution impact on human health, and the approaches proposed in the work can be used to solve similar problems in other regions of China.

1.5 Problems and Shortcomings

Applications of ground-based data and satellite remote-sensing technologies significantly increase the ability to monitor urban air pollution, but there are also many problems and shortcomings:

1. Air pollution remote-sensing monitoring accuracy is not high enough. In the process of remote-sensing monitoring of air pollution, the accuracy of different algorithms is increasing. As the remote-sensing inversion algorithm cannot meet the accuracy requirements, the harm of air pollution is difficult to quantify the description, its medical, health, and other limited fields of application.
2. Lack of atmospheric vertical information. Satellite remote-sensing data contains the comprehensive information about the atmosphere; when the information on atmospheric air pollution is extracted from these remote-sensing data, atmospheric vertical information is often needed.
3. The temporal resolution of satellite remote-sensing data is not high enough. The temporal resolution of satellite remote-sensing data is mostly better, so it is not enough to monitor air pollution accidents and most of the air pollutants.
4. Because of the wide regional and regional differences in pollution, the regional characteristics of pollutants are relatively fragmented. The spatial and temporal distribution of air pollution in different regions and the causes of pollution meteorological are lack of comprehensive understanding.
5. Some biological mechanisms have shown that changes in pollutant concentrations and changes in meteorological factors have a negative impact on human health; in some degree, meteorological elements determine the distribution and diffusion of pollutants.
6. Actually, there are relatively few studies on the statistics and mechanism of the changes of pollutant concentration and temperature change and its impact on human health at home and abroad. The research on sensitive disease prediction based on pollutants, pollution meteorological parameters, and meteorological factors is also rare.
7. $PM_{2.5}$ and PM_{10} related research focused on urban areas, background, and rural areas of the relative lack of research. Pollution is more serious in urban areas of pollutants that can be transported to the impact of the lower wind in direction of the region or even the entire region. So, the observations in the background area help to understand the pollution of the entire region.

1.6 Book Outline

In short, this study has great significance to the improvement of social economy and the quality of people's life. At the same time, it can also help to solve similar problems in other parts of China.

The full text is divided into seven chapters; chapter's arrangement and the main research contents are the next:

Chapter 1: Introduction. Introduction describes the background of the research, the present situation, and the significance of the research at home and abroad.

Chapter 2: Study Area, methodology, and data. Firstly, this chapter describes the basic situation of Lanzhou City, and then it describes the research methods used in this paper. Finally, the selection and processing of the research data are explained.

Chapter 3: This chapter contains an analysis of the AOD and the UHI during 2010–2016 basing on the images obtained from the Landsat and MODIS satellite (Terra/Aqua), using an algorithm of the LST and NDVI, to determine the urban heat island effect, to determine the atmospheric pollution, and to estimate the environmental situation in the city of Lanzhou. In order to facilitate the analysis of typical dust weather conditions of atmospheric aerosol total backscattering coefficient, aerosol subtype images and absorbing aerosol index in Lanzhou were selected for the period from April 24 to April 27, 2014. The CALIPSO instrument data analysis was conducted in conjunction with an analysis of the system and forecasting NAAPS (Navy Aerosol Analysis and Prediction System) aerosol models and the characteristics of the vertical distribution of the aerosol and dust, as well as with the OMI (Ozone Monitoring Instrument) which was used to study the areas affected by the sandstorm.

Chapter 4: Study on air pollution in Lanzhou City from 2003 to 2012. Firstly, this chapter evaluates the air quality of Lanzhou City in the last 10 years. Secondly, this paper determines SO_2, NO_2, PM_{10}, and dust as the main research indicators of air pollutants. The annual variation, seasonal variation, and diurnal variation of concentrations of air pollutants in Lanzhou were analyzed by Spearman's rank correlation coefficient. Finally, the correlation between concentrations of SO_2, NO_2, and PM_{10} in Lanzhou City and the meteorological data of Lanzhou City are analyzed, and the correlation between concentrations of SO_2, NO_2, and PM_{10} and meteorological parameters are further studied.

Chapter 5: Analysis of air pollutant concentrations and sources in Lanzhou City. This chapter mainly studies the total pollution of Lanzhou City and the characteristics of concentration of air pollutants in Lanzhou City from 2013 to 2016. Firstly, the variation characteristics of total air pollution in Lanzhou are obtained through the analysis of air quality index (AQI) in Lanzhou. Secondly, the seasonal variation characteristics of concentrations of the main air pollutants in Lanzhou are studied through the systematic analysis of the concentration and correlation of SO_2, O_3, and NO_2 in the air of Lanzhou during 2013–2016. Finally, the main sources of air pollution in Lanzhou are tracked, and the correlations between concentrations of $PM_{2.5}$ and PM_{10} in different seasons are studied.

Chapter 6: Analysis of the causes of air pollution in Lanzhou and its influencing factors. In this chapter, the classification of dust storms affecting Lanzhou and its surrounding areas is studied, and the sources of dust storms and their impacts on regional air quality are analyzed. In addition, the meteorological conditions affecting air pollution are also studied in this chapter.

Chapter 7: Conclusion and prospect

In summary, this chapter is based on different seasons and source multiscale data of aerosol lidar, used here are MODIS, CALIPSO, OMI, Landsat 8, ground air quality monitoring station, etc.; methods are based on synergetic analysis, on mathematical statistics, on HYSPLIT, and on NAAPS. The main sources of air pollutants in Lanzhou, the evolution mechanism of air pollution, and the seasonal variation of air pollutants are systematically studied, and the expected research objectives are estimated.

Chapter 2
Location and Methods of Investigation

2.1 Location of the Study

2.1.1 Geographic Location

Lanzhou (N36°02′ and E103°48′) is located in Northwestern China. It is the capital of Gansu Province. Its population is 3.68 million people (in 2016). The total area is 13085.6 km^2. It is located in the upper reaches of the Yellow River, near the border between the Tibetan Plateau, the Loess Plateau, and the Inner Mongolia Plateau (Fig. 2.1). Lanzhou is an industrial city with a developed chemical, oil, and flower industry, manufacturing of heavy equipment and coal, and production of electricity from plants. Lanzhou has a major railroad, road, and air hub for the whole Northwestern China. For this reason, Lanzhou became an important strategic and trading center on the Silk Road in ancient times.

2.1.2 Characteristics of the City Districts

The sub-provincial city of Lanzhou has direct jurisdiction over five districts and three county-level cities. General characteristics of the population, area, and population density of the districts are presented in Table 2.1.

Xigu is the westernmost region and the main industrial region of Lanzhou with large petrochemical plants, power plants, and other heavy industries, including the aluminum smelting and production of textiles. The eastern part of Lanzhou Chengguan is a large residential area where there are a coal-fired power plant and a wide variety of agricultural, engineering, and pharmaceutical industries, research institutes, and the main railway station. In the Qilihe District, southeasterly of Xigu, manufacturing industries including petrochemical, machine-building, textile, electrical, locomotive, plastic, and food industries are located. Anning, located northerly

M. Filonchyk, H. Yan, *Urban Air Pollution Monitoring by Ground-Based Stations and Satellite Data*, https://doi.org/10.1007/978-3-319-78045-0_2

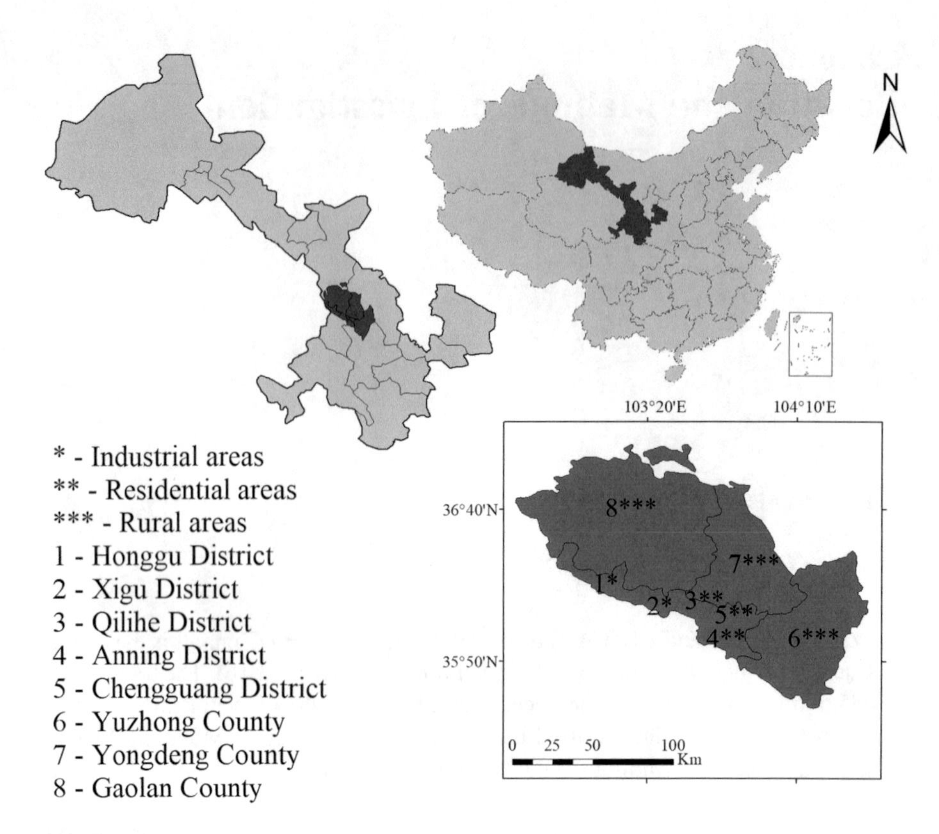

Fig. 2.1 Geographic location of the studied area

Table 2.1 The characteristics of administrative division of Lanzhou City

Name	Population(2010 census)	Area (km^2)	Density(km^2)
Chengguan District	1,278,745	220	5812.47
Qilihe District	561,020	397	1413.14
Xigu District	364,050	385	945.58
Anning District	288,510	86	3354.76
Honggu District	136,101	575	236.69
Yongdeng County	418,789	6090	68.76
Gaolan County	131,785	2556	51.55
Yuzhong County	437,163	3362	130.03

of Xigu and Qilihe on the opposite bank of the Yellow River, is mainly a residential city where the main activity is agriculture and education.

The city is located on a bend in the narrow river valley, and due to that it is hemmed in with no free air flow. Therefore, the discharge of industrial and domestic pollutants, the long narrow river valley location, and presence of unvegetated surfaces nearby are combined to ensure ideal conditions for high levels of dust deposition.

2.1.3 Climate and Environmental Problems

Lanzhou has a temperate continental climate. The average annual temperature is 10.3 °C. Lanzhou has cool summer and warm winter and a famous summer resort. The average annual sunshine hours are 2446 h, frost-free period is 180 days, and the average annual rainfall mainly concentrated in June–September is 327 mm.

Gansu is located in the northwest inland territory where there is no sea, air temperature and humidity do not easily reach there, and rains are rare, so most areas have dry and strong continental temperate monsoon climate. It is characterized by long cold winter, uncertain spring and summer lines, short summer, high temperature, and short cool autumn. The annual average temperature in the province is 0–16 °C, and the temperature difference is big; the sunshine is sufficient. Throughout the province, precipitation is from 36.6 to 734.9 mm.

At the rapid development of the economy and society in Lanzhou City, energy demand continues to increase; a substantial increase in the number of motor vehicles is a frequent regional air pollution phenomenon, restricting the sustainable development of society and economy and affecting people's health. Air pollution prevention and control work situation are grim; facing the pressure is unprecedented and long term. There are many reasons for the air pollution in Lanzhou City; one of them is the special geography landform. The first one is Lanzhou which is located in the Loess Plateau valley terrain in a semi-closed huge dumbbell obviously. The city is located in the basin, the North-South confrontation between two mountains; the relative elevation is of 660 m; it extends about 2–8 km from north to south, and it extends about 35 km from east to west; the basin topography caused by air pollution is not easy to spread out. The second one is the influence of meteorological factors. The average annual static wind rate is of 62.7%, the static wind rate is more than of 80% in winter, and the average daily wind speed is less than of 0.8 m/s in Lanzhou City. Winter stable weather accounted for more than 70% of stable stratification and easy formation of inversion layer. The third one is the industrial and energy structure; the reason is not reasonable. The forth one is the existence of coal pollution; the reason is serious. Currently, there are 170,000 households in urban and rural areas surrounding the city with small stove and 200 sets of vertical heating boiler, there are a considerable number of street stalls, and the structural pollution is particularly prominent in the heating period of coal. The fifth one is the increasing pollution of motor vehicles. In 2013, there were more than 600,000 motor vehicles in Lanzhou; in recent years, the average annual growth rate is more than 10%. Due to poor urban traffic and low speed or speed of motor vehicles, exhaust emissions increased, and the main road formed a significant pollution belt, and the proportion of nitrogen oxides in the air was accelerating in an upward trend. The sixth one is a serious dust pollution. In recent years, due to accelerating the pace of city construction and efforts increasing the city surrounding mountain reclamation engineering cut a large area of the building caused by dust pollution is becoming increasingly serious, and the prevention and management are difficult. The seventh one is the public awareness of environmental protection which needs to be further improved.

2.2 Sources of Data

2.2.1 Ground-Based PM and Gaseous Pollutants Concentration Data

In order to protect air quality and to monitor air quality standards, the Ministry of Environmental Protection of the People's Republic of China (MEP) and Gansu Provincial Department of Environmental Protection constructed more than five air quality monitoring stations in the Lanzhou City illustrated in Fig. 2.2 and Table 2.2. The ground-based monitoring stations measure the concentration of weighted particles in the air and show hourly and daily average concentration. The mass concentrations of $PM_{2.5}$, PM_{10}, SO_2, NO_2, CO, and O_3 were recorded by the Data Center, the Ministry of Environmental Protection of the People's Republic of China (http://datacenter.mep.gov.cn/ and http://113.108.142.147:20035/emcpublish/) and Data Center Gansu Provincial Department of Environmental Protection (http://61.178.220.12:8016/).

Fig. 2.2 Distribution map of environmental air quality monitoring points in Lanzhou City

Table 2.2 Coordinates of air quality monitoring stations in Lanzhou City

Observation spots	Latitude	Longitude	City districts
Lanlian hotel	36°06′05.2″	103°37′57.9″	Xigu
Worker hospital	36°04′28.1″	103°42′50.3″	Qilihe
Biological Product Institute	36°04′17.2″	103°50′36.6″	Chengguan
Railway Design Institute	36°02′23.3″	103°50′21.4″	Chengguan
Lanzhou University Yuzhong Campus	35°56′32.4″	104°09′31.2″	Yuzhong

2.2.2 Ground-Based Meteorological Measurements

The ground-based measured meteorological data obtained from many Chinese standard stations are available online from National Meteorological Information Center, China Meteorological Administration (http://www.nmc.cn). The data obtained from many stations are open to the public, including a daily, 10-day, monthly, and annual meteorological elements. The main measurements are surface pressure, atmospheric temperature, humidity, wind, precipitation, sunshine hour, cloud, etc.

The ground-based meteorological parameters, surface wind speed (WS), temperature (T), and relative humidity (RH) were used in this book.

2.2.3 Data of MODIS

Acquisition and pretreatment of MODIS (moderate-resolution imaging spectroradiometer) satellite data is a new generation of optical remote-sensing instrument in the world. At the same time, as MODIS has the advantages of wide band and high temporal and the US National Aeronautics and Space Administration (NASA) on the implementation of MODIS data in the world conducts free policy, the Terra satellite remote-sensing image data on the MODIS were mainly used in this paper.

MODIS, the moderate-resolution imaging spectroradiometer, was mounted on two satellites, Terra and Aqua. It has spectral bands with 36 moderate resolution levels range from 0.4 to 14.4 μm (Table 2.3).

MODIS data products in various stages of processing are used to get different levels of products from the most original instrument data Level 0 to contain the statistical analysis of results of the Level 4, as shown in Table 2.4.

2.2.4 Data of Landsat 8

Landsat 8 is an American satellite of remote sensing, the eighth one under the Landsat program (the seventh one put into orbit). It was originally called as the Landsat Data Continuity Mission (LDCM) and created by NASA and United States Geological Survey (USGS). It was launched into the orbit on February 11, 2013. The satellite was built on the basis of the LEOStar-3 platform by a company named as Orbital Sciences Corporation. The payload of the spacecraft was designed by Ball Aerospace and Goddard Space Flight (NASA); the launch was made by United Launch Alliance. Approximately 100 days after the withdrawal LDCM, configuration and inspection were held under the control of NASA. On May 30, 2013, after the completion of inspections, LDCM was placed under the management of USGS and received the official designation of Landsat 8.

Table 2.3 Channel, band width, and basic purpose of MODIS

Band	Wavelength(nm)	Resolution(m)	Primary use	Band	Wavelength(nm)	Resolution(m)	Primary use
1	620–670	250	Land/cloud/aerosols boundaries	20	3.660–3.840	1000	Surface/cloud temperature
2	841–876	250		21	3.929–3.989	1000	
3	459–479	500	Land/cloud/aerosols properties	22	3.929–3.989	1000	
4	545–565	500		23	4.020–4.080	1000	
5	1230–1250	500		24	4.433–4.498	1000	Atmospheric temperature
6	1628–1652	500		25	4.482–4.549	1000	
7	2105–2155	500		26	1.360–1.390	1000	Cirrus clouds water vapor
8	405–420	500		27	6.535–6.895	1000	
9	438–448	1000	Ocean color/phytoplankton/biogeochemistry	28	7.175–7.475	1000	
10	483–493	1000		29	8.400–8.700	1000	Cloud properties
11	526–536	1000		30	9.580–9.880	1000	Ozone
12	546–556	1000		31	10.780–11.280	1000	Surface/cloud temperature
13	662–672	1000		32	11.770–12.270	1000	
14	673–683	1000		33	13.185–13.485	1000	Cloud Top Altitude
15	743–753	1000		34	13.485–13.785	1000	
16	862–877	1000		35	13.785–14.085	1000	
17	890–920	1000	Atmospheric water vapor	36	14.085–14.385	1000	
18	931–941	1000					
19	915–965	1000					

Table 2.4 MODIS data level

Level	Description
Level 0	The reconstructed full-resolution raw data without the overhead sensor (synchronization frames, headers, repetition)
Level 1A	The reconstructed full-resolution raw data of the sensor, linked in time, provided with radiometric and geometric calibration coefficients and parameters of geo-referenced (attached to this level required for correct calibration factors are calculated according to Level 0, but not used at this level)
Level 1B	The radiation or radiation flux density produced by the calibration algorithm in the Level 1A data
Level 2	Derived geophysical variables (height of ocean waves, soil moisture, ice concentration) with the same resolution as that of the data of Level 1
Level 2G	With Level 2 products are similar but contains a pixel to the grid mapping
Level 3	The data of earth physical parameters, which are average and grid, are corrected and compounded in time and space
Level 4	Data obtained from calculations based on previous levels. The final model or previous analysis result of processing layers (variables resulting from a plurality of measurements)

Table 2.5 Operational Land Imager and Thermal Infrared Sensor spectral bands

Band	Resolution (m)	Wavelength (μm)	Description	Sensor
1	30	0.433–0.453	Coastal/aerosol	Operational Land Imager
2	30	0.450–0.515	Blue	Operational Land Imager
3	30	0.525–0.600	Green	Operational Land Imager
4	30	0.630–0.680	Red	Operational Land Imager
5	30	0.845–0.885	Near infrared	Operational Land Imager
6	60	1.560–1.660	Short-wave infrared	Operational Land Imager
7	30	2.100–2.300	Short-wave infrared	Operational Land Imager
8	15	1.360–1.390	Panchromatic	Operational Land Imager
9	30	0.52–0.90	Cirrus	Operational Land Imager
10	100	10.6–11.19	Thermal infrared	Thermal Infrared Sensor
11	100	11.5–12.51	Thermal infrared	Thermal Infrared Sensor

Landsat 8 uses two sets of tools, i.e., the Operational Land Imager (OLI) and the Thermal Infrared Sensor (TIRS). The first set obtains images in nine bands of visible light and near infrared (IR); the second set obtains images in two ranges of far (thermal) IR (Table 2.5). The satellite was designed for active lifetime in 5.25 years, but the amount of fuel you can use was designed up to 10 years.

2.3 Methodology

2.3.1 Criteria for Evaluation of Air Pollution

Ambient air quality has been regulated in China since 1982. In China in 1982, initial limits were set for TSP (total suspended particles), SO_2, NO_2, lead, and BaP (Benzo(*a*) pyrene). In 1996, the standard was both strengthened and expanded from 1982 levels under National Standard GB3095–1996 (Table 2.6).

Due to the fact that this book covered the period from 2003 to 2016 for research from 2003 to 2012, NAAQS would be used, because at the time it was relevant GB3095–1996.

In February 2012, China's Ministry of Environmental Protection (MEP) released a new Chinese National Ambient Air Quality Standard (NAAQS) GB3095–2012 (that replaced and reviewed GB3095–1996), which set limits for $PM_{2.5}$, PM_{10}, SO_2, NO_2, CO, and O_3. The new standards took a nationwide effect in 2016. Table 2.7 contents the air quality standards in China, as specified in GB3095–2012.

In parallel with GB3095–2012, a new air quality index (AQI) definition was also released. The new AQI, which includes both $PM_{2.5}$ and ozone for the first time, is specified in HJ633–2012. It replaces the old Air Pollution Index (API); the API had been used till 2012 and was replaced with a new one. The air quality index is a kind of reflection and evaluation of air quality; there are several air pollutant concentrations of routine monitoring simplified as a single numerical form, an indirect method to characterize the degree of air pollution and fractionation of air pollution and air quality situation, and its characteristic are comprehensive, simple, intuitive, and suitable for the city air quality status and trends; it is commonly used in the comprehensive evaluation of the atmospheric environmental quality index internationally, using this index form. The AQI is the highest value calculated for each pollutant as follows (EPA, 2016):

Table 2.6 National Ambient Air Quality Standard (as specified in GB 3095–1996)

		Limit			
Pollutant	Collecting time	Class 1	Class 2	Class 3	Unit
SO_2	Annual	0.02	0.06	0.10	mg/m^3
	24 h	0.05	0.15	0.25	
	Hourly	0.15	050	0.70	
NO_2	Annual	0.07	0.08	0.08	
	24 h	0.08	0.12	0.12	
	Hourly	0.12	0.24	0.24	
PM_{10}	Annual	0.04	0.10	0.15	
	24 h	0.05	0.15	0.25	

Table 2.7 National Ambient Air Quality Standard (GB3095–2012)

Pollutant	Averaging time	Limit		Unit
		Class 1	Class 2	
SO_2	Annual	20	60	μg/m^3
	24 h	50	150	
	Hourly	150	500	
NO_2	Annual	40	40	
	24 h	80	80	
	Hourly	200	200	
O_3	Daily, 8 h maximum	100	160	
	Hourly	160	200	
PM_{10}	Annual	40	70	
	24 h	50	150	
$PM_{2.5}$	Annual	15	35	
	24 h	35	75	
CO	24 h	4	4	mg/m^3
	Hourly	10	10	

Table 2.8 Breakpoints for the AQI

O_3 (ppb) 8 h	O_3 (ppb) 1 h[a]	$PM_{2.5}$ (μg/m^3) 24 h	PM_{10} (μg/m^3) 24 h	CO (ppm) 8 h	SO_2 (ppb) 1 h	NO_2 (ppb) 1 h	AQI
0–54	–	0.0–12.0	0–54	0.0–4.4	0–35	0–53	0–50
55–70	–	12.1–35.4	55–154	4.5–9.4	36–75	54–100	51–100
71–85	125–164	35.5–55.4	155–254	9.5–12.4	76–185	101–360	101–150
86–105	165–204	55.5–150.4	255–354	12.5–15.4	(186–304)[b]	361–649	151–200
106–200	205–404	150.5–250.4	355–424	15.5–30.4	(305–604)[b]	650–1249	201–300
[c]	405–504	250.5–350.4	425–504	30.5–40.4	(605–804)[b]	1250–1649	301–400
[c]	505–604	350.5–500.4	505–604	40.5–50.4	(805–1004)[b]	1650–2049	401–500

[a]Usually AQI is calculated basing on the 8-h ozone concentrations. However, there are few areas where an AQI based on 1-h ozone values would be more precautionary. In these cases, in addition to calculating the 8-h ozone index value, the 1-h ozone value may be calculated, and the maximum of the two values may be reported

[b]One-hour SO_2 values do not define higher AQI values (≥200). AQI values of 200 or greater are calculated

[c]Eight-hour ozone values do not determine the higher AQI values (≥301). The AQI values from more than 301 are calculated basing on the 1-h concentration of O_3

Step 1: Using Table 2.8 to find the two determining the highest concentration of all observers in each region reporting and truncate as follows: Ozone (ppm) is truncated to three decimal places; $PM_{2.5}$ (μg/m^3) is truncated to one decimal place; PM_{10} (μg/m^3) is truncated to integer; CO (ppm) is truncated to one decimal place; SO_2 (ppb) is truncated to integer; NO_2 (ppb) is truncated to integer.

Table 2.9 Air quality index, air quality, and health implications (based on HJ663–2012)

AQI	Air quality index	Air quality	Health implications
0–50	1	Excellent	No air pollution
51–100	2	Good	Few hypersensitive individuals should reduce outdoor exercise
101–150	3	Lightly polluted	Slight irritations may occur, individuals with breathing or heart problems should reduce outdoor exercise
151–200	4	Moderately polluted	Almost everyone's health will be affected; the adverse effects on the sensitive population is particularly evident
201–300	5	Heavily polluted	Everyone's health will be affected more seriously
>300	6	Severely polluted	Everyone's health will be the most seriously affected

Step 2: Breakpoints that contain a concentration.
Step 3: Using the formula to calculate the AQI:

$$I = \frac{I_{\text{high}} - I_{\text{low}}}{C_{\text{high}} - C_{\text{low}}}\left(C - C_{\text{low}}\right) + I_{\text{low}} \tag{2.1}$$

where I = the (air quality) index; C = the pollutant concentration; C_{low} = the concentration breakpoint that is $\leq C$; C_{high} = the concentration breakpoint that is $\geq C$; I_{low} = the index breakpoint corresponding to C_{low}; I_{high} = the index breakpoint corresponding to C_{high}.

Table 2.9 gives the air quality index range and corresponding air quality categories; the long-term exposure to air pollution index exceeding 200 would pose a health scare for people; the greater the AQI index, the greater the harm to human body is. In a word, since the implementation of the quantitative characterization of the status of air pollution and the public health, the current AQI tips play a great role in promoting the environmental protection work, especially open and transparent, and play a positive role in improving the public awareness of environmental protection.

2.3.2 Satellite-Based Measurements

The Hybrid Single Particle Lagrangian Integrated Trajectory Model (HYSPLIT)

The Hybrid Single Particle Lagrangian Integrated Trajectory Model (HYSPLIT) is a complete system for the calculation of simple areas of air trajectories and complex transport, dispersion, chemical transformation, and deposition modeling. It was developed by the NOAA (National Oceanic and Atmospheric Administration) and Australia's Bureau of Meteorology. The HYSPLIT becomes one of the most widely used models of atmospheric transport and dispersion in the atmospheric sciences community. The HYSPLIT has been used in different models describing atmospheric transport, dispersion, and deposition of pollutants and hazardous materials. Some examples of applications include the tracking and forecasting of radioactive material of forest fires, dust, smoke, wind pollutants from different stationary and mobile sources of emissions, allergens, and volcanic ash (Stein et al. 2016; Rolph et al. 2017).

The dispersion of the pollutant was calculated by assuming either tightening or dispersing the particles. In the layered model, the puffs expand until they exceed the cell size of the meteorological cell (horizontally or vertically) and then divide into several new puffs; each of them has its own mass fraction of pollutant. In the particle model, the number of particles associated with the model region, the mean wind field, and the turbulent component were fixed. The default configuration of the model assumes a three-dimensional distribution of particles (horizontal and vertical).

Cloud-Aerosol Lidar and Infrared Pathfinder Satellite Observations (CALIPSO)

The Cloud-Aerosol Lidar and Infrared Pathfinder Satellite Observations (CALIPSO) uses an innovative lidar and visualizing system and is intended to make 3D images of clouds and atmospheric aerosols, as well as to study the formation and "composition" of clouds and the way they influence the weather, climate, air quality, and rains. This task is made by the CALIPSO polarizing laser radar. It allows to differentiate between water vapor and ice crystals as well as between liquid and solid particles of atmospheric aerosols. The CALIPSO vertical profile data range from 2 km below the world ocean level to 40 km above sea level with the highest lidar resolution of about 0.33 km horizontally and about 30 m vertically at an altitude from −0.5 to 8.2 km; at upper atmosphere, at the altitude exceeding 8.2 km, the horizontal resolution of CALIPSO is about 1 km and vertical resolution 60 m (Hunt et al. 2009; Winker et al. 2010).

Ozone Monitoring Instrument (OMI)

The OMI is installed in Aura satellite launched in July 2004, with a large territorial coverage (2600 km) and high spatial resolution (13 × 24 km). The instrument provides daily data collection from almost all over the world. The Earth is viewed in 740 wavelength bands along the satellite track with a swath large enough to provide global coverage in 14 orbits (1 day). As the OMI can observe the reflected data surface ranging from visible (350–500 nm) to the ultraviolet (UV–1, 217–340 nm, UV–2, 306–380 nm) band, it can monitor the UV absorptive aerosols and distinguish their types, such as dust and biomass combustion aerosols (Zhang and Tang, 2012). The OMI's UV absorbing aerosol index (AAI) is a key parameter to identify dust aerosols by comparing the two ultraviolet bands (354, 388 nm), calculating the radiation transmission and considering that the output power of molecular scattering used to estimate the global area of dust aerosol content. The total AAI is positive for the corresponding UV absorbing aerosol, the AAI is negative for the nonabsorbing aerosol. As the cloud top albedo and the wavelength of light do not almost correlate, the AAI may distinguish properly between aerosols and clouds, and the AAI of clouds and other nonabsorbent aerosols are very small with their values being about 0 (Torres et al. 2007a, b, 2013). Dust aerosol and biomass combustion aerosol are high-value strong absorptive aerosols, so the use of AAI can be more precise as compared to remote sensing of dust storm weather, development, and their transition mechanism.

Navy Aerosol Analysis and Prediction System (NAAPS)

The Navy Aerosol Analysis and Prediction System (NAAPS, NRL-Navy Research Laboratory, http://www.nrlmry.navy.mil/aerosol) is a real-time operational system to forecast the global distribution of tropospheric sulfate, dust, and smoke. The NAAPS is a modified form of a sulfate aerosols model developed by Christensen (1997). Its operation involves the global meteorological fields from the Navy Operational Global Atmospheric Prediction System (NOGAPS) (Hogan and Brody 1993). NAAPS is intended to simulate and predict the concentration of aerosols, dust aerosols, sulfate, and smoke aerosols on a global scale. The vertical resolution of the pattern has 24 layers (from the ground to 100 hPa height), the horizontal resolution is $1° \times 1°$, and the simulated time interval is 6 h.

2.3.3 Method of Extracting Ecological Factors of Underlying Surface

The urban heat-island effect is highly correlated with the composition and change of the underlying structure of the city, as well as of the analysis of the impact factors of urban heat island, and the urban heat-island effect is necessary for the underlying surface of the ecological factor information. Therefore, according to the research needs and the correlation between ecological factors and heat-island effect, the urban heat-island effect is necessary to obtain the land surface temperature (LST), the normalized difference vegetation index (NDVI), and the aerosol optical depth (AOD).

Normalized Difference Vegetation Index

The NDVI (normalized difference vegetation index) or a relative normalized vegetation index is a simple quantitative measure of the amount of photosynthetically active biomass (it is also usually called as vegetation index). It is one of the most common and used indexes to solve problems, using quantitative assessment of vegetation.

The NDVI (normalized difference vegetation index) is calculated using the formula:

$$\mathrm{NDVI} = \frac{\mathrm{NIR} - \mathrm{RED}}{\mathrm{NIR} + \mathrm{RED}} \tag{2.2}$$

where NIR means reflections in the near-infrared region of the spectrum, and RED means reflected in the red region of the spectrum.

According to the formula, the density of vegetation (NDVI) to a certain point of the image is equal to the difference of reflected light intensities in the red and infrared wavelengths divided by the sum of their intensities.

To display the NDVI, standardized continuous gradient or discrete scale was used, showing values between –1 and 1% or in the so-called scaled range from 0 to 255 (used to display some packages of the Earth remote sensing (ERS) processing, corresponding to the number of gray shades) or in the range 0…200 (–100…100), which is more convenient, since each unit corresponds to 1% change in the indicator. Due to peculiarities of the reflection in the NIR-RED regions of the spectrum, natural objects are not related to vegetation and have a fixed value of the NDVI, which allows to use this parameter to identify them (Table 2.10).

Land Surface Temperature

A mask LST (land surface temperature) is used to calculate the Earth's surface temperature index, used in different climate studies.

1. MODIS Land Surface Temperature Products

   The level-3 MODIS global land surface temperature (LST) and the emissivity 8-day data were composed from the daily 1 km LST product (MOD11A1) and stored on 1 km. A sinusoidal grid shows the average values of clear-sky

Table 2.10 Identification of the underlying surface on the value of NDVI (values of NDVI are for photosynthetically inactive objects)

Object type	Reflection in the red region of the spectrum	The reflection in the infrared region of the spectrum	Value of NDVI
Dense vegetation	0.1	0.5	0.7
Sparse vegetation	0.1	0.3	0.5
Open soil	0.25	0.3	0.025
Clouds	0.25	0.25	0
Snow and ice	0.375	0.35	–0.05
Water	0.02	0.01	–0.25
Artificial materials (asphalt, concrete)	0.3	0.1	–0.5

Table 2.11 Calibration constants for calculating the temperature

Range	
K_1 (band 10)	774.8853
K_1 (band11)	480.8883
K_2 (band 10)	1321.0789
K_2 (band 11)	1201.1442

LSTs during an 8-day period. Calculation of the mask was based on the "Generalized Split-window LST" algorithm, described in the MODIS Land-Surface Temperature Algorithm Theoretical Basis Document (ATBD-MOD-11). Emission values recorded in 31 and 32 channels; converted to the corresponding temperature values, latitude for initial temperature approaches and humidity, zenith angle of sensor, cloudiness value, and the presence of water; and calculated for the respective pixel exceptions were used as input values. The temperature correction for all pixels used the average emission value (except for the pixels identified as having the snow cover, it used their own values for them).

2. Conversion of Radiance to At-Sensor Temperature from the Satellite Landsat 8
According to the values of the thermal channels, it was possible to define brightness temperature of the underlying surface. The accuracy of temperature estimation was theoretically about 0.5 °C; however, the haze in the atmosphere reduced the value to a few degrees.
The TIRS band data can be converted from spectral radiance to brightness temperature using the thermal constants provided in the metadata file:

$$T = \frac{K_2}{\ln\left(\frac{K_1}{L_Y} + 1\right)} - 273.15^{\circ}\mathrm{C} \tag{2.3}$$

where K_1 and K_2 are a band-specific thermal conversion constant from the metadata (K1_CONSTANT_BAND_x, where x is the band number equal to 10 or 11) (Table 2.11) and L_Y is a spectral radiation, which comes on the satellite sensor.

The temperature values are defined by the Landsat 8 channels 10 and 11 (they are differed via heat range covered intervals), and they differ from each other at 1.5–3 °C.

Aerosol Optical Depth

Since the launch of Terra (EOS-AM1) satellite in 1999 and launch of Aqua (EOS-PM1) in 2002 with MODIS instrument onboard, it had been possible to obtain data in a wide wavelength range from 0.4 to 14 μm. Firstly, the presence of the blue spectral range (0.47 μm) in MODIS system allowed to obtain the AOD data over much of the land surface. The algorithm for determining the aerosol optical thickness uses No. 1–7 and No. 20 MODIS data channels, as well as mask contours of cloud cover. In order to study the features of aerosol fields distribution and the impact of man-made and natural factors (industrial activity, forest fires, etc.) over the territory of Lanzhou City and Gansu Province, the MODIS instrument data used AOD data at a wavelength of 0.55 μm ($\tau_{0.55}$), products MOD04_L2 (Aerosol), and MOD08_D3 (Level 3 Daily Joint Aerosol/Water Vapor/Cloud Product), obtained from archiving system and dissemination data of the LAADS Web (Level 1 and Atmosphere Archive and Distribution System) obtained for the period 2005–2016, on the basis of Deep Blue algorithm (Hsu et al., 2013). The Deep Blue algorithm of Collection 6 covers the whole Earth surface including dark and light zones (e.g., such dust sources as the Taklamakan and Gobi deserts). The Deep Blue algorithm was first developed to obtain aerosol data and aerosol properties over deserts and other arid areas and further was improved to cover the earth surface with vegetation (Lee et al., 2016). The Deep Blue algorithm facilitates the use of relatively low reflection ratio of desert and other areas in the deep blue range to learn the aerosol properties. Both results are given in a 10 × 10 pixel scale (10 km by ground nadir). A presumed error of the Deep Blue algorithm from the Collection 6 is significantly lower than that of Collection 5.1 (Sayer et al. 2015a, b). Data, used in the study, was obtained during the midday observations of MODIS using the Deep Blue algorithm, because it was necessary to get AOD over lighter parts of a land.

Scientists called it as measurement aerosol optical thickness. It was a measure of how much light the airborne particles prevented from traveling through the atmosphere. Aerosols absorb and scatter incoming sunlight, thus reducing visibility and increasing optical thickness. Optical thickness of less than 0.1 indicated a clear sky and maximum visibility, whereas Fig. 2.1 indicated the presence of aerosols in such an amount that it would be difficult for people to see the sun, even in the afternoon. Table 2.12 shows what conditions the respective AOD values (Levy et al. 2010a, b) conform to.

Table 2.12 AOD Levels and the conforming atmospheric conditions

AOD	Atmospheric conditions
0	Clear sky, free from aerosols
0.02	Very clean
0.2	Fairly clean air
0.6	Air polluted
1	Heavy smoke/dust event

2.3.4 *Statistical Analysis*

After atmospheric environmental monitoring results, the most direct way to inspect the quality of atmospheric environment is to evaluate the concentration of each pollutant. The short-term pollution status of each pollutant can be analyzed by using the concentration of a pollutant, and the long-term variation law of each pollutant can be analyzed statistically.

Basing on the evaluation of the concentration, the pollution characteristics of each pollutant are mainly studied through a statistical analysis. General in the actual assessment, the main project is the statistical estimation of average value (average daily, monthly, and yearly mean), moving average, median, standard deviation, variance, kurtosis, skewness, minimum value, maximum value, etc. The daily average and annual average pollutant concentration of the statistical results are compared with national standards, and the pollution can be known according to the exceeding multiples. Other statistic results, such as the median and standard deviation, variance, kurtosis, skewness, and the most value, can determine the distribution of pollutant concentration, known as the general condition of pollution. Study on the change rule of pollutants during the year (season, month, and day), we usually draw a dynamic trend graph of pollutant concentration over time (year, quarter, month, and day), such as a line chart and a histogram. Basing on the dynamic change trend of the pollutant concentration, it can be very good to judge the pollution situation of the pollutants in different time periods and the change of the time.

Chapter 3
Research of Aerosol Optical Depth and Urban Heart Island in Lanzhou City by Means of Earth Remote Sensing

This chapter contains an analysis of the aerosol optical depth (AOD) and the urban heat island (UHI) basing on the images obtained from the Landsat and MODIS satellite (Terra/Aqua), using software ENVI5.1 and ArcGis10.2 and using an algorithm of the land surface temperature (LST) and normalized difference vegetation index (NDVI), to determine the urban heat island effect and the atmospheric pollution, to estimate the environmental situation in the city of Lanzhou completely, and to perform analysis and forecasting models of aerosol according to the NAAPS (Navy Aerosol Analysis and Prediction System), as well as to determine the characteristics of the vertical distribution of aerosols and dust using CALIPSO.

3.1 Spatial and Temporal Variability of the Temperature of Earth's Surface in Lanzhou City

The distribution of relative temperature within the city and its surroundings was considered by the example of images, obtained by MODIS systems (Terra satellite) and Landsat 8. The main objective was to study the spatial and temporal dynamics of the internal structure of Lanzhou surface heat island on satellite images in the thermal infrared range.

The images obtained by the spectrometer MODIS (Fig. 3.1) in the thermal channel were interesting because they generally describe the distribution of thermal radiation on the Lanzhou territory and its surroundings. The situation clearly showed that the city appeared as a heat island in the thermal image, namely, it stood out by high brightness in comparison to the surrounding area. The dark area in the northwestern part of the image is cloudiness. Due to the low resolution of image within the city, it was a problem to isolate specific objects, associated with thermal anomalies. Despite the lack of spatial resolution for detailed study of urban areas, MODIS thermal images had important virtue: it was possible to obtain absolute values of thermodynamic temperatures of the surface some times a day.

M. Filonchyk, H. Yan, *Urban Air Pollution Monitoring by Ground-Based Stations and Satellite Data*, https://doi.org/10.1007/978-3-319-78045-0_3

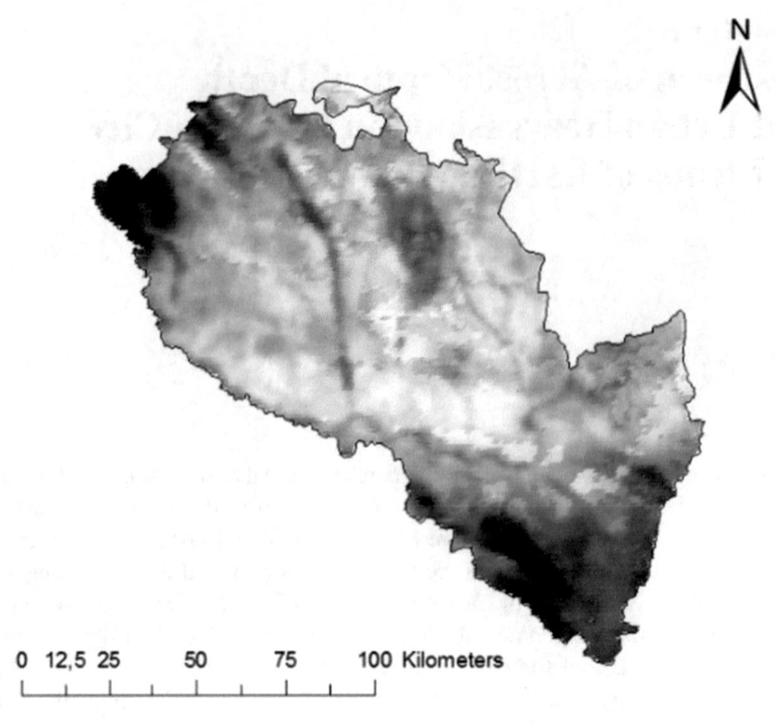

Fig. 3.1 A fragment of the thermal image of the territory of administrative division of Lanzhou City, obtained by MODIS system (MOD11A2) on May 26, 2010

The Landsat 8 image (Fig. 3.2) is characterized by a spatial resolution of about eight times higher than MODIS/Terra; the city facilities can be identified, focusing on the well-marked characteristic bend of the Huang He River and the mountains around the city. The resolution of the image allows selecting thermal anomalies of the city and linking them with the specific urban objects. In comparison with the MODIS/Terra image, the TM/Landsat 8 image was well deciphered: there were the river and its thermal contrasts, the forest vegetation, and the largest industrial enterprises.

The intensity of the thermal radiation depended on the emission of the object and was known as a function of the thermal radiation physical properties and the temperature of the radiating surface. Mapping, giving an idea of the distribution of the temperature of physical surface in Lanzhou, is based on images of Landsat 8, and its surroundings, based on MODIS images.

It was known that the ability of heat radiation of objects was not only marked as a diurnal variation, but it was also dependent on the season of the year. In mid-latitudes, where there are different types of weather, the total amount of incoming solar radiation and the nature of the atmospheric circulation typical to different seasons, these features, have their effect on the formation of urban heat island. Seasonal differences illustrated in the image of city objects and areas adjacent to the

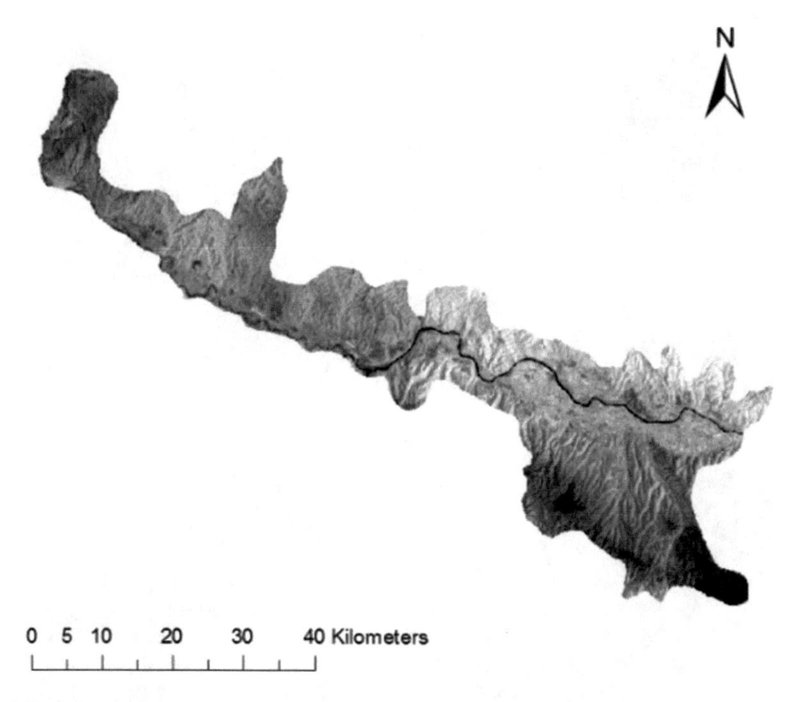

Fig. 3.2 A fragment of the thermal image of the central part of Lanzhou City obtained by Landsat 8 on July 16, 2016

city, as the example of comparing images of different seasons, are received by MODIS and Landsat 8 systems in the thermal range.

According to calculations, on February 2, 2015, within the analyzed landfill covering Lanzhou and the surrounding areas (Fig. 3.3), the territorial range of variability of surface temperature was equal from about −9 to +10 °C. As expected, the structure of the thermal field was not uniform. A surface temperature in the major part of the city was of +6…+7 °C; at the same time, extensive spots were fixed in the central part of the city with temperatures above +3 °C. Surface areas with different intensities of heat radiation were well highlighted. That was the obvious contribution of the industrial zone of the city into anthropogenically caused thermal anomalies. In consequence of that, the temperature of the Earth's surface in the industrial area was of +6…+10 °C.

As the image data obtained from MODIS showed, on April 15, 2015, the temperature field of the Earth's surface fluctuated between +10 and +34 °C. Spatial variability within the city and its surroundings was characterized by a variety of contrasts. In the winter period, the surface temperature in the industrial area was higher than in residential areas. So, in industrial and agricultural areas of Yongdeng Country and Honggu District, the surface temperature was of +29…+34 °C. In urban areas, where there were city's residential areas, the temperature was of +15…+26 °C.

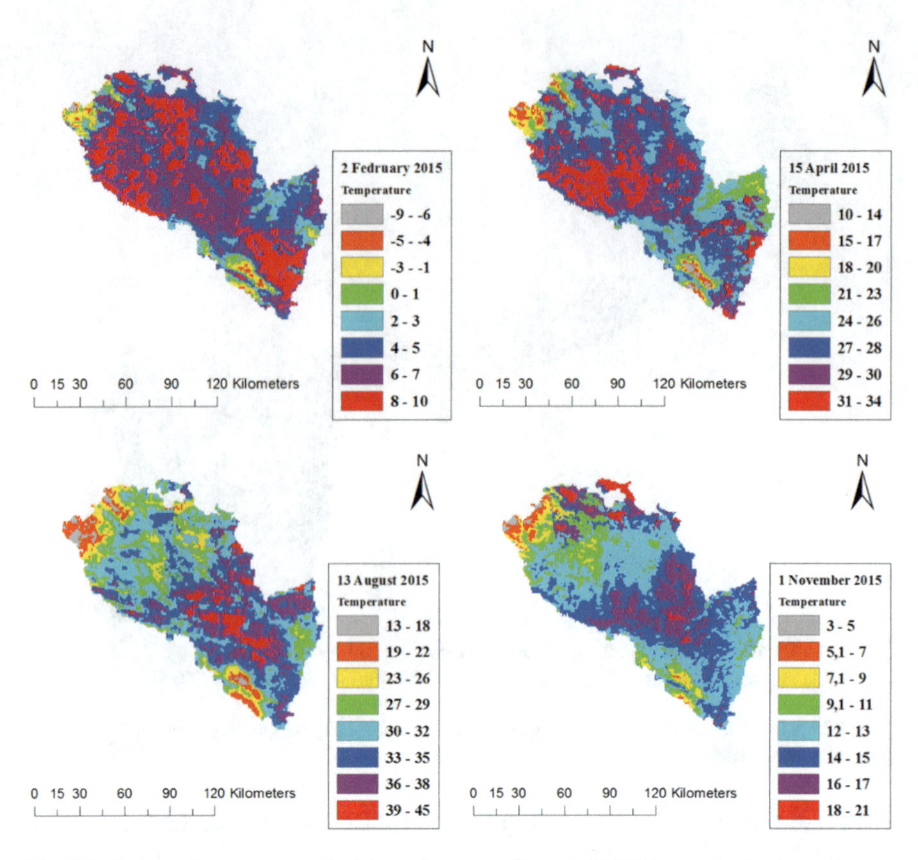

Fig. 3.3 Estimated temperature of urban surface according to MODIS images in the thermal range

On August 13, 2015, the territorial variability of the thermal field was characterized by the values in the range of +13…+45 °C (Fig. 3.3). The spatial homogeneity of the intensity of the thermal radiation outside of the city (mainly in the agricultural land of the city), as well as in the winter, was noticeably higher. In comparison to winter conditions (at comparable temperature contrasts), the structure of the investigated field had significant differences. Temperature anomaly of the city in summer mainly acquired the form of small spots, commensurate with the individual municipal facilities. The maximum temperatures of surface in the summer were of +36…+45 °C and were observed on the hills surrounding the city.

Considering the image of November 1, 2015, the dynamics of the distribution of the surface temperature of the Earth can be traced similar to other seasons (Fig. 3.3). As other images illustrated, the high temperatures were observed in industrial areas of the city with a temperature of 16+…+21 °C. In residential areas the temperature ranges from +5 to +13 °C. In general, the surface temperature was of +3…+21 °C.

The high-resolution images were used to detect temperature anomalies in urban areas. Figure 3.4 illustrated the images in the thermal infrared range of Landsat 8 on the territory of Lanzhou on the following dates: January 21, 2015; May 7, 2015;

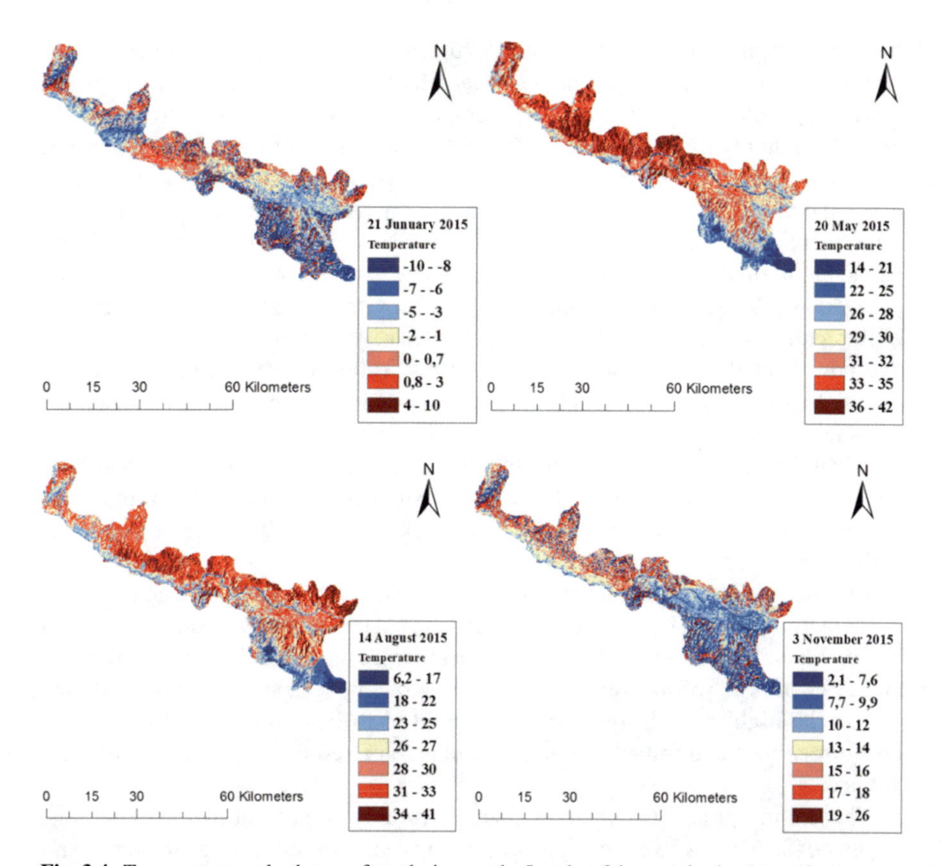

Fig. 3.4 Temperature and urban surface design on the Landsat 8 images in the thermal range

August 14, 2015; and November 3, 2015. The fragments of images represented the appearance of heat islands in every season. Snapshot of May 7 represented here a spring situation, because it was very difficult to find pictures of cloudless in spring months, so the picture for that date represented clouds.

In general, features of the distribution of thermal contrasts in the snapshots data can be summarized as follows:

The autumn image (Fig. 3.4) represents distribution of brightness, corresponding to differences in the intensity of thermal radiation of the Earth's surface, and it differs in contrast compared to other pictures. Water with high heat capacity accumulates heat in the summer best of all allocated water bodies (rivers, lakes, etc.). Water accumulates heat over the summer, and by the end of autumn, when the thermal radiation of soil, vegetation and other objects decreases, water objects retain an increased level of radiation. At the same time, even small water objects, as well as waterlogged areas, stand out, which demonstrates the possibility to use such images to study soil moisture.

The winter image (Fig. 3.4) is characterized by a clear allocation of residential development, some water objects, and separate enterprises. Winter was a specific time of the year for heat island in cities in the middle latitudes. Sometimes, foreign researchers argue that the problem of heat islands is not so acute for cities where winters are cold, and even on the contrary, this phenomenon allows to reduce the cost of heating buildings. However, in winter, the impact of the heat island can be felt not less strongly, than in the summer, although it manifests itself a little differently: the number of cloudy days dramatically increases, the amount of precipitations increases, and air pressure is lowered over the city. As for mapping heat island on thermal images, winter was the only season, when many industrial zones radiated heat weaker than residential areas. That was due to the active functioning of the central heating in residential houses. Elevated level of thermal radiation heat and power plants was associated with central heating operation, illustrated on a picture by the bright spots with browse steam plume. In winter, it was easy to find the water objects, which were the most heavily exposed by thermal pollution. They stood out by an increased level of radiation, in comparison with other water bodies. This was because the natural water underwent cooling in winter stronger, and the main source of heat, causing the temperature anomaly in water bodies, is anthropogenic heat.

In comparison with the autumn and winter images, the spring image (Fig. 3.4) is characterized by a heat-contrast image of the island. Here we can see the most "cold" water objects (they have not been warmed up enough after the winter), and a few more "warm" vegetation (active growing season has not been started yet, so the high thermal lows are still devoid of foliage forests, and parks have not formed yet) and industrial facilities, characterized by high intensity of radiant heat.

The summer image (Fig. 3.4) is characterized by significant thermal contrasts. Vegetation is in the active growth phase, and it is actively transpiring moisture evaporates, formed due to this powerful thermal lows. Furthermore, the foliage provides shading surface under trees, reducing its heat level. Urban buildings and structures actively accumulate and radiate heat, increased at the expense of the overall intensity of the heat island. Industrial zones form powerful thermal highs due to man-made release of heat during the production process and during the absorption and emission of solar radiation directed on the industrial buildings.

The map is drawn to determine the thermal structure of urban areas of the Lanzhou City (Fig. 3.5). The map was compiled on the basis of a multi-temporal image processing and composed of 25 different-season thermal images obtained from 2013 to 2016, while comparison with ultra-high resolution satellite images. Seven clusters are allocated there.

The map shows the differentiation of geosystems on the seasonal dynamics of the thermal radiation intensity. This differentiation appears differently depending on the geographic features of the area.

So, the zones 1 and 2 determine the location of several most heavily emitting anthropogenic geosystems (the largest industrial enterprises in the city), as well as natural tectonic objects with height of more than 1800 m. The zone 3 has anthropogenic geosystems, characterized by dense buildings and a minor amount of natural objects (vegetation). The zone 4 has natural and anthropogenic geosystems, which

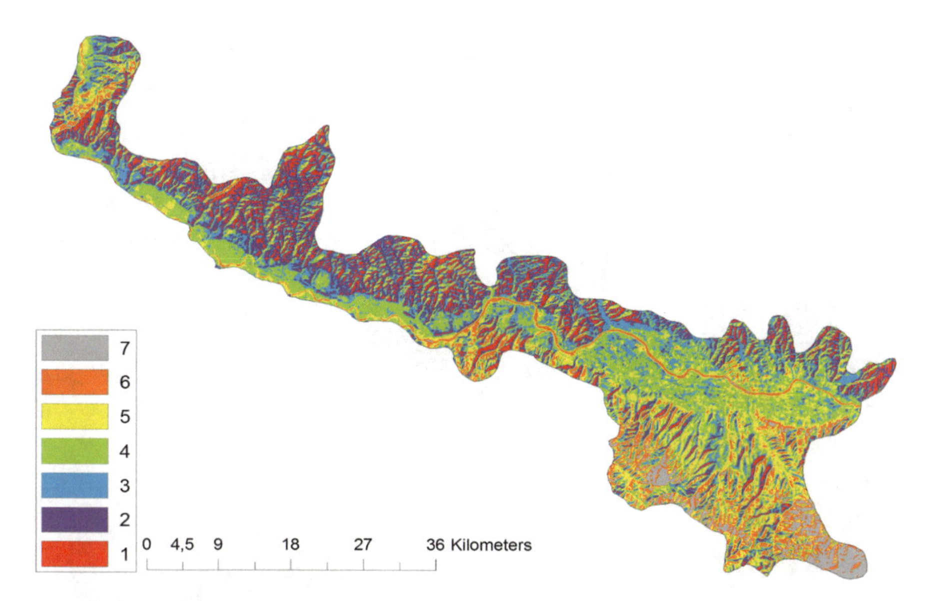

Fig. 3.5 The thermal structure of the Lanzhou City: 1 is the zone of extremely high intensity of thermal radiation; 2 is the zone of very high intensity of thermal radiation; 3 is the zone of high intensity of thermal radiation; 4 is the zone of average intensity of thermal radiation with significant seasonal amplitudes; 5 is the zone of average intensity of thermal radiation with reduced seasonal amplitudes; 6 is the zone of reduced intensity of thermal radiation; 7 is the zone of high intensity of thermal radiation in the cold season and low intensity in warm

are characterized by a significant role of anthropogenic factors, such as residential areas with a high degree of gardening, urban parks, arrays of garden areas, and bushes. The zones 5 and 6 have geosystems with forests and shrub arrays of natural and natural anthropogenic origin (parks, gardens).

Elements of the thermal structure of geosystems occur in certain study areas or their sets. The divide of the geosystems is based on the effect of lead for each of factors of the studying areas: for example, for urban and suburban areas, there is anthropogenic transformation of the area. For natural areas, leading factors of formation heterogeneity geosystems are relief (primarily, the exposure of slopes), vegetation cover, manifestations of volcanic activities, and moistening of areas. Consequently, the maps of thermal structure allow to reveal different geosystems of studying area, basing on leading factors of differentiation.

It should be noted that a high-altitude zone of vegetation appears on natural areas with difficult terrain (e.g., mountain systems) on the thermal structure maps.

In considering urban sites, they were the sources of thermal anomalies. According to Gartland (2012), surfaces of all the types can be divided into four groups depending on their daily peak temperature: (1) trees, grass, and vegetation are the coldest surfaces with maximum daytime temperatures, (2) anthropogenic light coatings (roads, sidewalks, parking, etc.) are warmer, (3) anthropogenic black or gray coating, and (4) roofs are the hottest surfaces in the cities and suburbs with a maximum daily temperature.

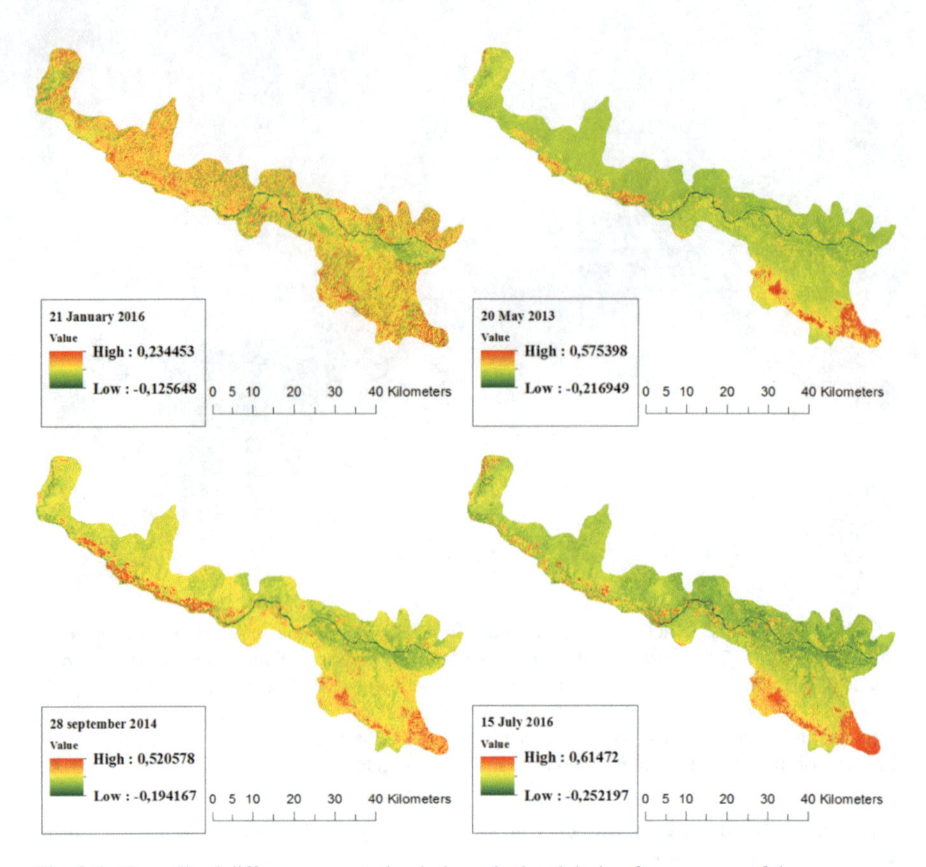

Fig. 3.6 Normalized difference vegetation index calculated during four seasons of the year

3.2 Research of Aerosol Optical Depth and Urban Heart Island in Lanzhou City by Means of Earth Remote Sensing

The direct effect of green space on the formation of the thermal field in the summer can be estimated by the distribution map of the NDVI (Fig. 3.6) basing on the Landsat 8 data (wavelength 10.6–11.19 μm and 11.5–12.51 μm) of Lanzhou.

The main objective was to determine the current state of the city ecosystems, to estimate their diversity, and to analyze the productivity of vegetation and its anthropogenic transformation degree using the vegetation indices.

Many researchers paid attention to studying the correlation between the vegetation capacity and the thermal radiation intensity if comparing the thermal infrared images and characteristics of the normalized difference vegetation index, based on the data of red and near-infrared channels (NDVI); they always found a pronounced feedback. This correlation was often assessed by image analysis of LST and of NDVI (Pu et al. 2006; Tan et al. 2010). Using the NDVI and LST, additional information of the Earth's surface was obtained, especially the information on evapotranspiration and soil moisture (especially in areas with sparse vegetation). The results of the combined analysis of the LST and of the NDVI images were often used to assess the impact of urbanization (Weng 2009), and an additional using of the LST and NDVI maps and land cover/land use described the situation in the best way (Tan et al. 2010). However, they often denoted the lack of efficacy of the NDVI 34 as a means of evaluating the effect of vegetation in the city. Other methods of assessing the impact of urbanization, such as collaborative analysis of LST and analysis of the vegetation fractions from spectral decomposition model, were offered (Weng et al. 2004).

The index can take values from −1 to 1. The vegetation NDVI can take positive values, typically from 0.2 to 0.9. When shoot occurs (during the vegetative period), the growth of vegetation biomass corresponds to an increase in NDVI, and with the onset of the maturation period, the chlorophyll content and, correspondingly, the NDVI value decrease.

It should be noted that the small woody vegetation areas, where NDVI (0.8–0.95) is high, contribute to a significant cooling of the physical surface in summer. Nevertheless, it can be noted that the areas of such lands in the city are not presented, and therefore they do not respond to the concepts of "urban ecological framework." Anyway, the cumulative effect of green space on thermal field in Lanzhou should be recognized as very limited effect. In considering seasonal index changes, seasonal index values varied from −0.125 to 0.234 in winter period, indicating a prevalence on vegetation over the whole city territory. High rates of the NDVI were recorded in summer, so, according to the Landsat 8 data, in the city area, the range was 0.252 to 0.614, noting a great influence of green planting in the city during a warm period of the year.

Generally, the analysis of obtained NDVI data (0 to 0.5) low vegetation was observed in all seasons. It was confirmed by geobotanical descriptions of key areas, which included mountain, desert, and urban research park areas. Despite a significant amount of biomass, the areas corresponded to the values of NDVI (0.02–0.3), which could be explained by the existence of large amounts of woody and withering stalks with a low content of chlorophyll. In this way, quantitative and qualitative criteria disturbance of plant communities as a result of anthropogenic impacts, identified by using the vegetation index, give an opportunity to make the plant communities more necessary for the long-term solutions to improve the natural communities.

3.3 Spatial and Temporal Variability of Aerosol Optical Depth Using the MODIS Data

As a result, the use of data obtained from the high-resolution Landsat cameras allowed to monitor the status of objects size up to a single field or forest stands. However, by using the MODIS/Terra data, characterized by large spatial coverage, areas could be monitored and commensurated with the areas of the regions and countries. Therefore, the whole territory of the Gansu Province and the entire administrative district of Lanzhou City, including both urban and rural areas and natural areas, were chosen to study the distribution patterns of atmosphere aerosols.

Aerosol pollution of the atmosphere is caused by man-made emissions and forest fires, affecting the chemical composition of the atmosphere. The necessity to study aerosols is caused by insufficient knowledge of aerosol particles effect on the atmospheric processes in Lanzhou and their contribution to the effect of global warming.

In order to monitor the distribution of aerosol fields over a vast territory of all Northwestern China and territory of Lanzhou City, the satellite monitoring method was compared to the economic and effective method of doing a range of ground station observations.

To explain the problem on a regional scale (Fig. 3.7), the average distribution of the AOD ($\tau_{0.55}$) was shown in China between 2005 and 2015. The spatial distribution of the AOD was characterized by two centers of the low AOD values and two centers of the high AOD values. The two low AOD values of the center were located in areas with high-vegetation cover and low density of population, namely, in the provinces of Inner Mongolia and Heilongjiang, and the largest natural flattering area in North China where the AOD indicators were 0.15–0.3. Also in three provinces of Yunnan, Sichuan, and Tibet in Southwest China, the AOD was 0.05–0.2. The two low AOD centers were connected by a low AOD zone (0.2–0.3) in a northeast-southwest direction across China. Compared to the spread of the centers with low AOD values, the centers with high AOD levels of indicators in a range of 0.7–0.9 were located on the territories of Yangtze River from Sichuan Basin, the territories of North China Plain, and the territories of central provinces of Hunan and Hubei to Yangtze River Delta and Pearl River Delta in Southern China, which were the most industrialized and densely populated areas of China. In the Tarim Basin region of the Xinjiang Province of Northwestern China, the territories were characterized by natural aerosols, which could be formed from the dust emitted from the Taklamakan Desert in the Tarim Basin.

Figure 3.8 illustrated the 10-year average AOD value ($\tau_{0.55}$) for winter, spring, summer, and autumn in the Mainland China. The AOD characteristics ($\tau_{0.55}$) were qualified in two space centers with high AOD values ($\tau_{0.55}$) and in two centers with low AOD values ($\tau_{0.55}$), which were not changed in all four seasons and changed a lot from spring to winter. Significant seasonal variations of AOD values ($\tau_{0.55}$) were high in the Mainland China, according to the seasonal changes in aerosol sources in Eastern and Northwestern China. Anthropogenic emissions of sulfate aerosols, pure and organic carbon in East China were connected to the seasonal cycle of human activities, including industry, agriculture and transport. Dust storms caused by natural aerosols appeared more often in Northwest China in spring (Zhao et al. 2006, 2013b; Zhi et al. 2007). The dust natural aerosol transport caused by spring dust storms could also enhance the AOD values in spring in other Chinese regions.

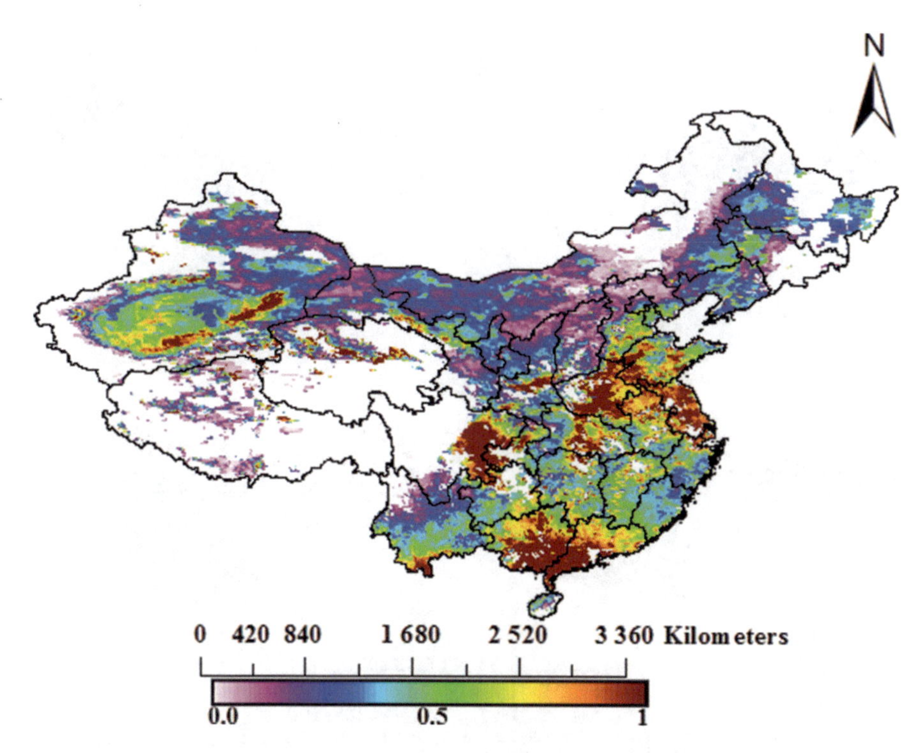

Fig. 3.7 Annual mean of the AOD ($\tau_{0.55}$) over Mainland China for 2005–2016

As illustrated in Fig. 3.8, the AOD during spring in the Central and Eastern China (excluding parts of Zhejiang and Fujian) and the Sichuan Basin were very high - between 0.6 and 1, among them, the Yangtze River Delta region, the middle reaches of the Yangtze River (Wuhan-Changsha Line) and the Sichuan Basin all reached a maximum of 1; AOD in the Pearl River Delta, eastern Anhui, Jiangsu, Henan, and southern Hebei were relatively high, between 0.7 and 0.9. In addition, there was also a high-value area of 0.9–1.0 in Xinjiang, and the AOD value of the Yellow River Basin was also about 0.4–0.5. The lowest AOD values in spring were still in the junction of Sichuan, in Yunnan and Qinghai-Tibet Plateaus, in the border between Tibet and Nepal, and in the northeastern Heilongjiang. Summer AOD values in the Sichuan Basin, the middle reaches of the Yangtze River, and the Pearl River Delta region were better than spring one. This might be related to the monsoon climate. Summer AOD value in the Sichuan Basin, in the middle reaches of the Yangtze River, and in the Pearl River Delta region is better than spring AOD value. This may be related to the monsoon climate. But in Jiangsu, Shandong (except Jiaodong Peninsula), Henan, Hebei, Tianjin, and Tangshan areas and in other places, the AOD value was relatively high, becoming the largest regional presence in China. The results showed that the high temperature in summer in North China could greatly improve the chemical conversion of two organic aerosols. In autumn, the AOD value in North China decreases significantly, which may be related to the increase of precipitation in

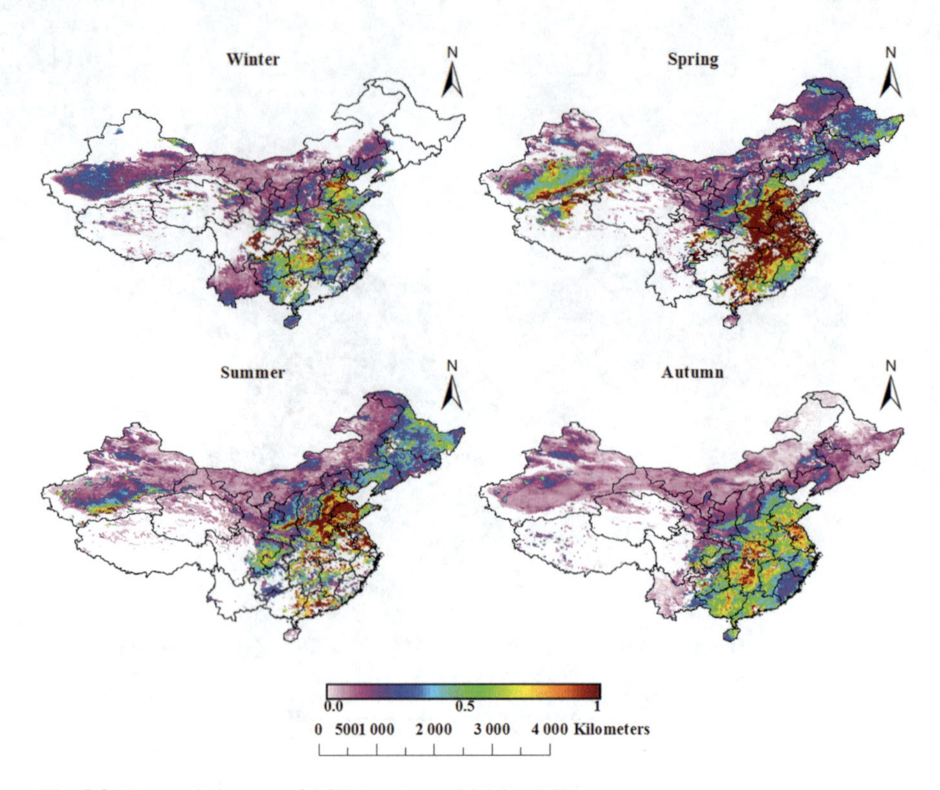

Fig. 3.8 Seasonal changes of AOD ($\tau_{0.55}$) over Mainland China

August and the increase of atmosphere aerosol wet deposition. At the same time, due to the end of the flood season in Sichuan Basin, the AOD value began to rise. In winter, the highest AOD value in China was in the Sichuan Basin, in the middle reaches of the Yangtze River, and in the Yangtze River Delta region. The eastern parts of Jiangxi, Guangxi, and Pearl River Delta were the sub-high-value areas. The lowest value appeared in Sichuan, Yunnan, and Qinghai-Tibet Plateau, and some areas were also low-value areas. In addition, the northeast and Inner Mongolia winter AOD values were also low.

According to Fig. 3.8, the seasonal average of AOD was 0.5, and it was defined as a high-value, depending on area, the maximum AOD value in China in spring, in summer, and in autumn was the largest, and the smallest one was in winter.

In general, the 10-year average statistics of AOD ($\tau_{0.55}$) on the territory of Mainland China indicated that in the most developed and large areas, AOD indicators were the following: Sichuan Basin (Chengdu, 0.815), North China Plain (Beijing, 0.591), Pearl River Delta (Guangzhou, 0.6), Yangtze River Delta (Nanjing, 0.721), Northwestern China (Xining, 0.215), Tibetan Plateau (Linzhi, 0.100), and Xinjiang (Urumqi, 0.423). The fact that the minimum AOD value was located on the territory of Tibetan Plateau, and the maximum one was on the territory of Sichuan Basin, which bordered Tibetan Plateau, indicating the importance of topography in the aerosol distribution throughout China.

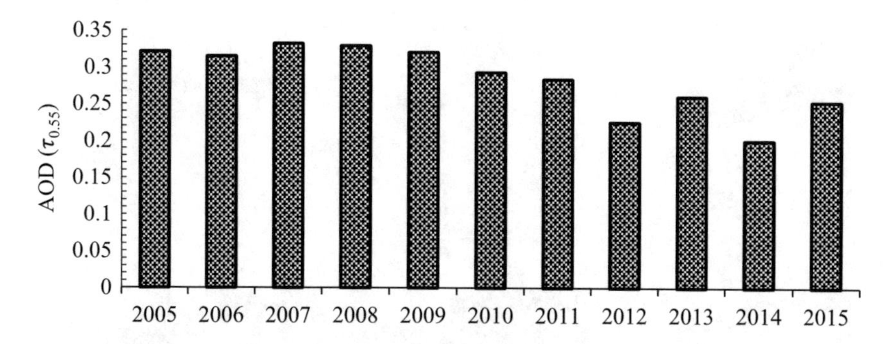

Fig. 3.9 Interannual variations in AOD in Lanzhou

To characterize the monthly variation of AOD, the area of Gansu Province and the city of Lanzhou were used as study areas.

Monthly data of AOD ($\tau_{0.55}$) as well as the MODIS data in the period of 2005–2016 was shown in Fig. 3.9. The results, illustrated in Fig. 3.9, indicate significant influence of aerosols on the formation of the ecological situation in the whole of Gansu Province, including one of the largest cities in western China – Lanzhou City. Almost throughout all territories of the province, high average AOD values were observed ($\tau_{0.55}$) in the spring, and relatively lower AOD values ($\tau_{0.55}$) were observed in autumn and winter months. The maximum values of AOD ($\tau_{0.55}$) were observed in most of the province in March and April with mean values of AOD ($\tau_{0.55}$) above 0.8, this indicates frequent sandstorms from the deserts of the North and Northwestern parts of China in the spring period (Zhao et al. 2006; Zhi et al. 2007). The low AOD values were similar, as well as with the entire territory of China, and were observed from September to January, with the testimony of 0.05 to 0.25.

Figure 3.9 illustrates that in Lanzhou City for the whole studied period, the value of AOD ($\tau_{0.55}$) was less than 0.35; the highest value was in 2007, when the AOD index ($\tau_{0.55}$) was 0.332; the lowest values were the same in 2014, when the indicators were 0.2. Generally, Fig. 3.10 shows a downward trend of atmospheric aerosol concentrations over Lanzhou.

In Lanzhou City, the maximum average was observed in April, with average values of 0.5 for 10 years; the minimum values of 0.05–0.15 were in October, November, January, and February. From March to July, the AOD average rose slightly in June, mainly due to the relatively abundant water vapor content in Lanzhou in June; the monthly average relative humidity was the highest in the year. In the case of higher relative humidity, aerosol hygroscopic growth can increase AOD significantly.

The spring in Lanzhou City was easily affected by the dust weather in the north or local pollution sources, the aerosol particles in the atmosphere were liable to increase the AOD in the spring, and the water content in the air increases with the increase of the temperature in the summer. The gas-particle transformation process of aerosols was accelerated under high-temperature and high-humidity conditions, increasing the ability of aerosol formation, so that the accumulation of urban aerosols produced effect.

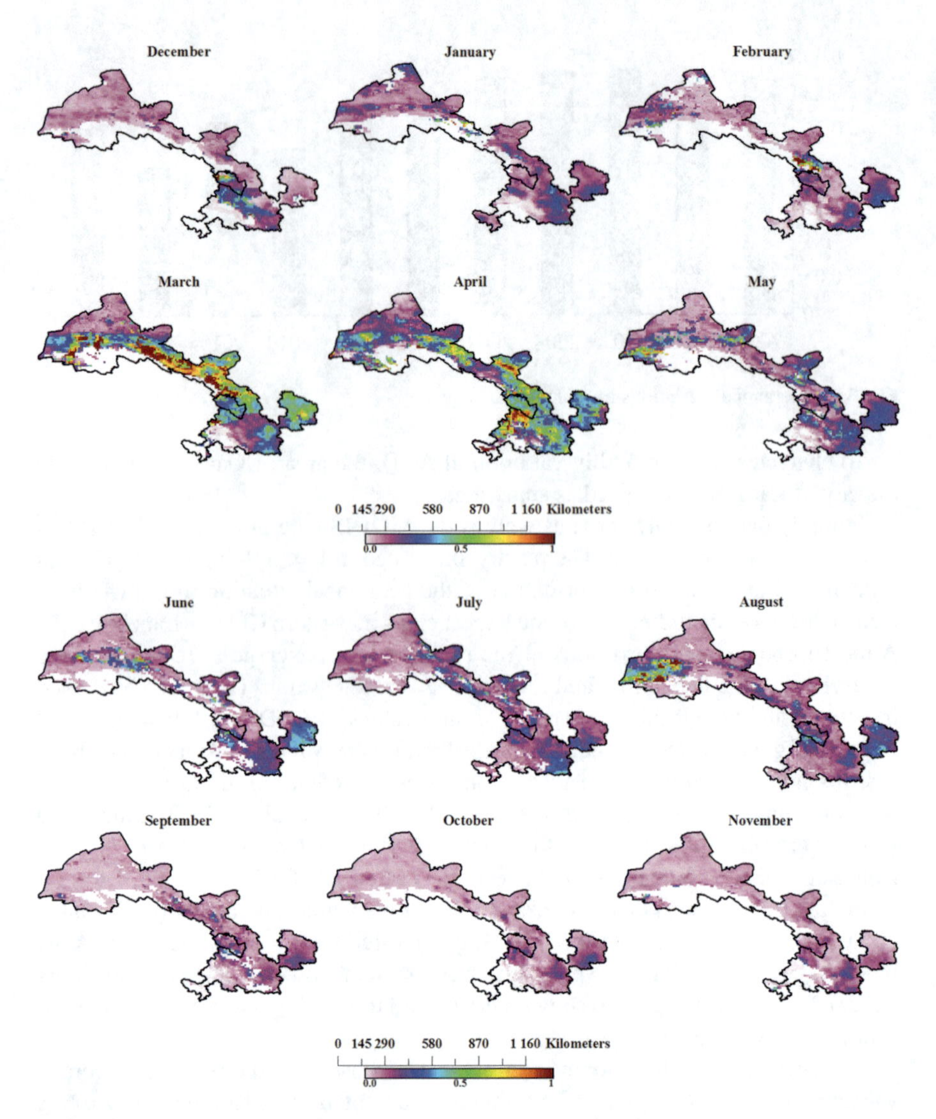

Fig. 3.10 Monthly variation of AOD ($\tau_{0.55}$), obtained from MODIS over Gansu Province for the period 2005–2015

In addition, the summer solar radiation was strong, atmospheric photochemical reaction was active, and it was conducive to the formation of secondary aerosols. Therefore, the hot weather conditions in summer were another important reason for the high AOD. However, the contaminants produced by the straw burned in the surrounding crops of the city would have a certain effect on AOD. Autumn and winter were mainly controlled by the mainland high-pressure system and when the atmospheric stratification was stable, frequency inversion temperature was high, pollutant diffusion conditions were poor, most of aerosols came from local anthropogenic sources, and concentration was relatively high, so that would lead to increased AOD. Especially, the haze weather was more serious period, when the AOD was significantly increased. One can see that the AOD value of Lanzhou City was at a high concentration level, and the seasonal variation was not obvious, which was mainly related to the weather situation and the influence of internal and external sources.

There were two main sources of aerosols in China: anthropogenic emissions in Eastern China where there was a high degree of industrialization, and population density, and natural and dust emissions in a desert in Northwestern China. The zone between these two main aerosol sources stretches from Northeast China to the Tibetan Plateau with good vegetation cover and the small population were dominant by the aerosols received from the aerosol sources, therefore it was necessary to consider the study of aerosol formation and migration based on regions rather than on a concrete territory.

3.4 Temporal and Spatial Distribution of Atmospheric Aerosol in Typical Dusty Weather in Lanzhou Using CALIPSO, OMI, and NAAPS

In order to facilitate the analysis of typical dust weather conditions of atmospheric aerosol total backscattering coefficient, aerosol subtype images and absorbing aerosol index in Lanzhou were selected from April 24 to April 27, 2014. The CALIPSO instrument data analysis was conducted in conjunction with an analysis of the system and forecasting NAAPS (Navy Aerosol Analysis and Prediction System) aerosol models and the characteristics of the vertical distribution of the aerosol and dust.

Between April 24 and 27, 2014, dust weather occurred in cities, as well as in the southern part of Xinjiang, Gansu, western Hexi Corridor near the Inner Mongolia, northern Ningxia, northern Shaanxi, northern Shanxi, and north of Northern China; as well as local sand dust storm events occurred in Hexi or Gansu Corridor and in the north of Northern China, being one of the largest dust weather processes which occurred in the north of China since spring 2014. According to the satellite data and ground monitoring station data, the comprehensive analysis revealed that the dust weather originated in the north and in the south of Republic of Mongolia and southern part of Xinjiang Basin in China. The sand and dust weather caused by strong cold air took place on April 24, 2014, in Gansu Hexi Corridor, Inner Mongolia, and the western region; in the morning of April 26, an extensive dust weather appeared in Gansu Hexi Corridor, Inner Mongolia Midwest, northern Ningxia, northern Shaanxi, and northern Shanxi.

Figure 3.11 illustrates the distribution of the absorbing aerosol index (AAI), observed by the OMI in Aura, in China on April 24–27, 2014. The figure indicates that the AAI has a larger value in southern Xinjiang, Hexi Corridor, northern Shaanxi, and northern Shanxi of China. The AAI indices in the southern Xinjiang

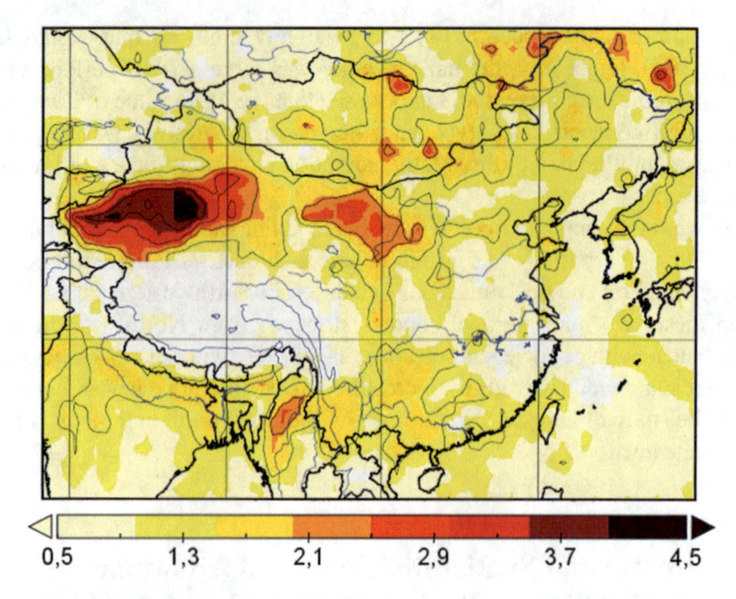

Fig. 3.11 Aerosol index obtained from OMI for dust storms over China: April 24–27, 2014

Basin and the northern part of the Northern China are especially obvious. The AAI value here was higher than 2.5 and that of Beijing and Tianjin was more than 1. The AAI index value is low in the northern part of Qinghai, and it is close to zero in the eastern part of Gansu. Figure 3.12 shows a rectangular area with the figure of vertical characteristics of the CALIPSO product within the same area; high-absorbing aerosols marked in this area are primarily dust aerosol, so the yellow color in the subtype classification indicates dust; the brown color indicates polluted dust. The paper shows the obvious dust weather occurred in five regions such as Xinjiang, Qinghai, Gansu, Shaanxi, and Ningxia. The paper analyzes the vertical distribution of dust aerosols in this region.

Based on the Navy Aerosol Analysis and Prediction System (NAAPS), a global forecast model predicting the concentrations of sulfate, dust, and smoke aerosols in the troposphere has been developed. Figure 3.13 shows that dust originated in Tarim Basin of southern Xinjiang and the central and southern regions of Mongolia on April 24 at 6 a.m. (UTC) when the combination of dust transporting processes in Inner Mongolia and in Gansu Province occurred. On April 25 at 6 a.m. in the source area, dust transported from western to the southern part of Xinjiang Province (Kashgar, Yarkant, Hotan) and to other regions; in addition, two dust branches transported to the east; under the combined action of two dust sources, the strength of sand dust affected Ningxia, Inner Mongolia, Gansu, and other locations. On April 26–27 at 6 a.m., in source region, two dust branches were transported in the southeastern direction, i.e., they were transported to China's Inner Mongolia, Hebei, Shanxi, Shaanxi, and other places.

As per above analysis, the dust process mainly originated in the southern Xinjiang Basin and in the central and southern regions of the Republic of Mongolia.

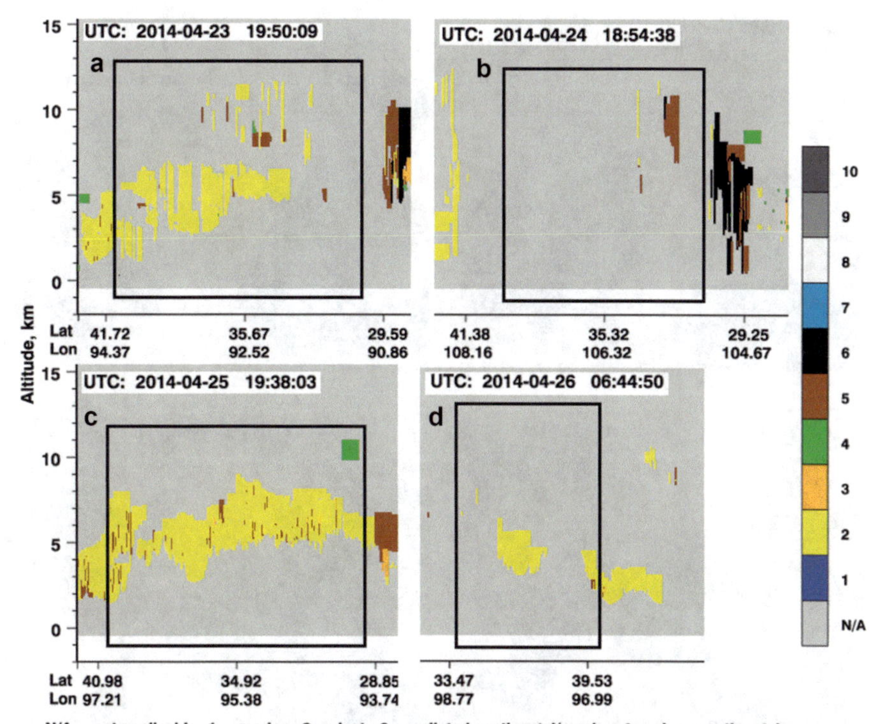

Fig. 3.12 Vertical feature mask obtained from CALIPSO for the dust storm. (**a**) April 23, 2014; (**b**) April 24, 2014; (**c**) April 25, 2014; (**d**) April 26, 2014

The transporting routes are the following: source area one (southern Xinjiang Basin), when dust was transported to affect the southwest of Xinjiang, Inner Mongolia, Gansu, Qinghai, Ningxia, and other regions; and source area two (southwestern Mongolia), where dust affected the central and western regions of Inner Mongolia, Ningxia, Gansu, northern Shaanxi, and northern Shanxi.

The CALIPSO tool was used for vertical dust analysis in dust weather. Based on the above analysis, we obtained the two paths of dust transporting, followed by the use of CALIPSO satellite to study the vertical distribution of the aerosol on April 23–27. Figure 3.14 shows data provided by CALIPSO satellite as of April 23 at 7:50 p.m. (UTC) which were observed near the dust source area (Inner Mongolia, southern Tarim Basin). On April 24 at 6:54 p.m., the satellite trajectory was observed in east Ningxia, west Shaanxi, and south Gansu. On April 25 at 7:38 p.m., the trajectory was located near the sources of aerosol contamination (Inner Mongolia, central and southern part of Tibetan Plateau) and near the dust source area (North Gansu and Hexi Corridor) on April 26 at 6:44 a.m.

Figure 3.14 shows the spatial distribution of the atmospheric backscatter factor of typical sand dust in the Northwestern China, which was calculated, interpolated,

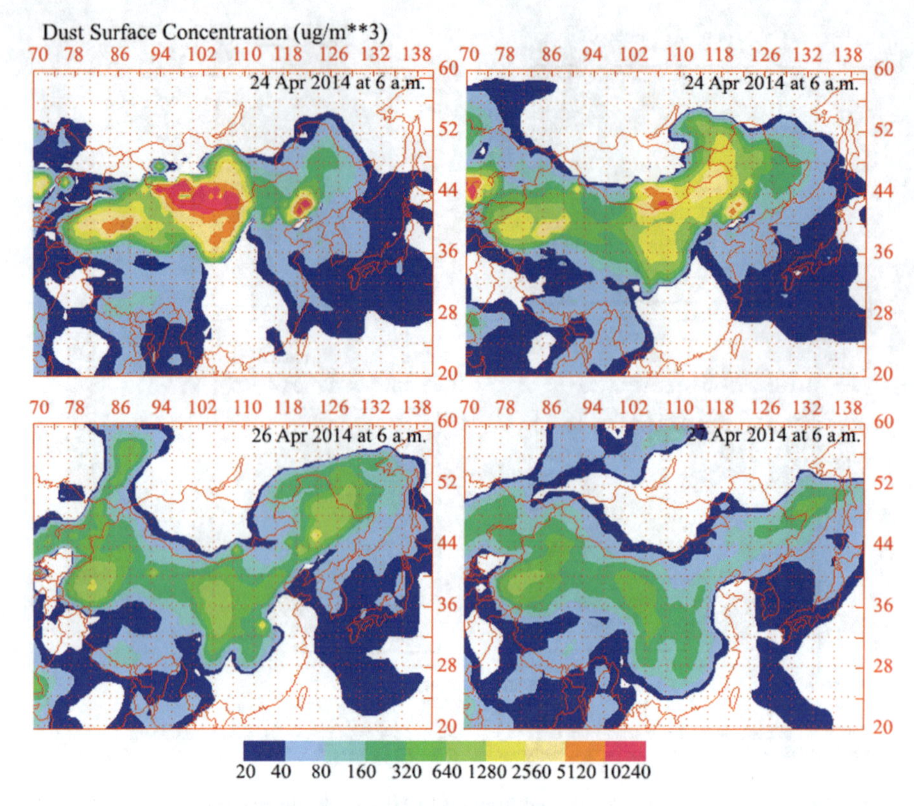

Fig. 3.13 Spatial distribution of dust concentrations from April 24–27, 2014, simulated using the Navy Aerosol Analysis and Prediction System (NAAPS) model

and averaged. The dusty weather conditions (Fig. 3.14) revealed the big amount of particles with their backscattering factor being 0.01~0.0001 $km^{-1} \times sr^{-1}$ and the aerosol distribution in the atmosphere within 0–12.5 km, indicating that aerosol particles proportion in the atmosphere was higher. The aerosol particles proportion gradually decreased from north to south due to its geographical distribution. It might be associated with higher humidity in upper atmosphere of the southern region featuring larger amount of ground vegetation as compared to sandy northern regions.

According to the two typical dust processes shown in Fig. 3.14 (April 23 and 24), the sand cloud was clearly divided into three parts. The highest cloud part was located at a height of about 12.5 km, where the upper layer mixture of clouds was located. Thus, dust aerosols may lift to middle troposphere height due to rise of the terrain from the aerosol source. Figure 3.14 (April 25 and 26) shows the division of the sand cloud into several parts, each part indicating the high and the low sand distribution. This is consistent with the change of altitude, i.e., the change of terrain altitude entails the change of sand clouds height. At the same time, a mixture of

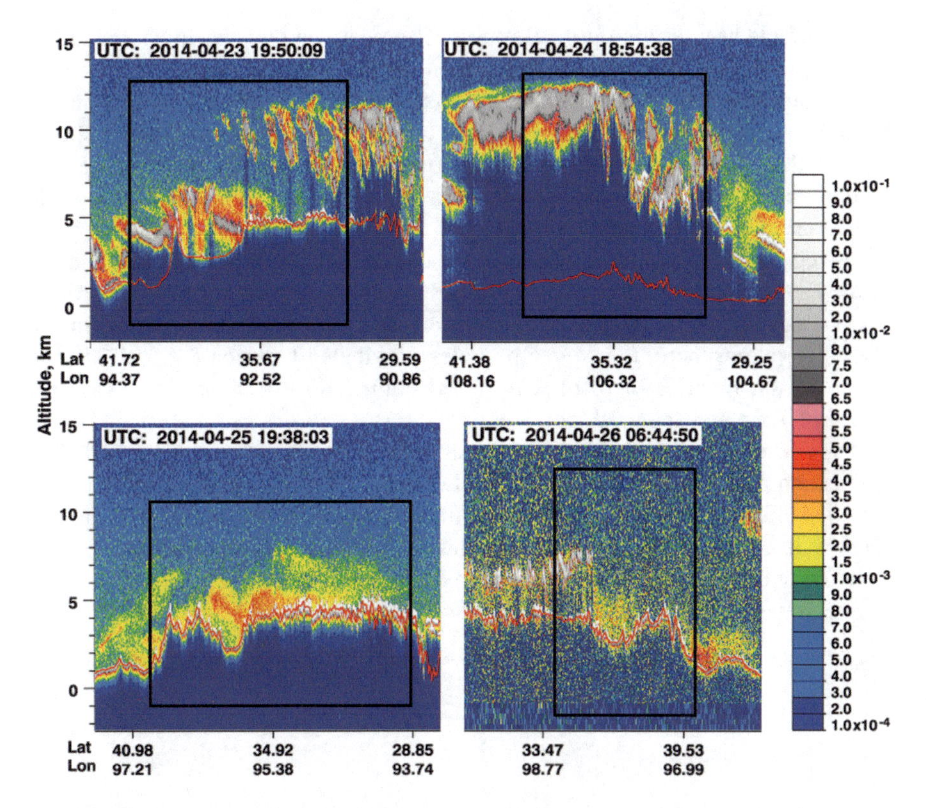

Fig. 3.14 Vertical distribution of total attenuated backscatter coefficients at 532 nm wavelength over Lanzhou District in dust storm weather from April 23–26, 2014

clouds occurs in the upper layers; the highest point of sand distribution at that period was located at about 8 km altitude. Yet, the larger part of the sand cloud was laid at about 5 km altitude. This might be associated with the rise of the terrain leeward side of the Tibetan Plateau, i.e., the sand cloud was significantly affected by the rise of the terrain. All the above results were obtained based on data on backscattering distribution character.

Basing on the distribution of China's UV aerosol index and the CALIPSO data obtained from the spaceborne lidar and detected by the ozone monitor, the regional and vertical distribution characteristics of the aerosol in the largest dust-free remote transportation event in 2010–2016 were carried out and studied, and the system analysis and forecasting models of aerosol by NAAPS were used. CALIPSO along with NAAPS can detect the distribution of dust aerosol effectively; sand cloud height and altitude changes show consistency. So, in the high-altitude area of the height of the sand clouds high and high-altitude areas of the upper reaches of the cloud mixture, the highest elevation of the sand was raised to the troposphere above the middle. That might occur due to the highest elevation of the area in the Helan Mountains leeward slope; the upper atmosphere was significantly affected by terrain uplift. Strong dust process had two main source areas, they were southern basins in China and Central and South Mongolia, and the process of dust transport was located in the southwest of Inner Mongolia and Gansu. Sand dust source one (southern basin) had conveyed the impact of China's southwestern Inner Mongolia, Gansu, Ningxia, and other regions; dust source two (south-central Mongolia) had affected the eastern part from Inner Mongolia, China, Gansu, North Shanxi, North Hebei, and Northeast China to western parts of the region. In sand and dust weather, the coarse particles were the main content of the near ground, and the coarse particles occupied the maximum value around 3 km. Irregular non-spherical aerosols increased with increasing height, in the high-level atmosphere, the coarse particles distributed between 0.4 and 2.0; the irregular non-spherical aerosol with a decoction ratio of 0.4 or more was at a height of 7 km and about 10 km.

The use of the spaceborne lidar for study aerosol optical properties had high application prospects. However, the aerial optical data of the spaceborne lidar system was affected by the satellite transit time; monitoring the diurnal variation in aerosols in a small area of studies was limited. Therefore, the use of remote-sensing technology for particle concentration monitoring is still necessary to carry out a lot of researches.

Chapter 4
Study of Air Pollution in Lanzhou City in 2003–2012

The data of the air pollutants used in this chapter are the monitoring data of the three SO_2, NO_2 and PM_{10} pollutants commonly used by the EPA. They are three pollutant indicators of monitoring data. According to the Chinese National Ambient Air Quality Standard (CNAAQS) GB3095–1996, as NAAQS GB3095–2012 was adopted in 2012, the study of the period 2003–2012 was using NAAQS GB3095–1996. The daily average concentration of major air pollutants of SO_2, NO_2, and PM_{10} and the air pollution index (API) during 2003–2012 were obtained from the Lanzhou Environmental Monitoring Center. Data on dust in the period 2003–2007 were obtained on the basis of monthly averages from automatic air quality monitoring stations located in the city. In addition, along with the "Lanzhou Environmental Quality Report" (2003–2007) and Ministry of Environmental Protection of the People's Republic of China, the data center published daily air quality data for key cities and other resources. The China Meteorological Administration Lanzhou ground weather station used meteorological data, including temperature data, air pressure data, 6 h precipitation data, wind speed data, visibility data, and dew-point temperature data, as well as it used the conventional surface meteorological observation data daily eight times at the same time.

4.1 Urban Air Quality Assessment

Lanzhou City has one of the most serious air pollution problems in our country and in the whole world according to the national city's daily air quality data released by the State Department of Environmental Protection (a number of key cities increased from 84 in 2005 to 86 in 2009, and it increased to 120 in 2011). The air pollution index in the city from 2003 to 2012 was focused on comparison, all days were compared with different pollution levels (due to different API values) and days when heavy pollution was observed (API> 300) were selected as comparison parameters.

M. Filonchyk, H. Yan, *Urban Air Pollution Monitoring by Ground-Based Stations and Satellite Data*, https://doi.org/10.1007/978-3-319-78045-0_4

Table 4.1 Grade of air quality distribution in Lanzhou in 2003–2012 (in days)

Year	Excellent	Good	Lightly polluted	Moderately polluted	Heavily polluted	Severely polluted
2003	11	196	87	45	15	11
2004	4	200	102	42	11	7
2005	15	223	77	27	10	13
2006	6	199	91	29	11	29
2007	27	244	63	20	7	4
2008	15	246	65	24	4	5
2009	10	226	98	21	3	5
2010	37	184	82	37	10	14
2011	16	226	78	22	8	15
2012	33	237	58	18	6	14

Table 4.2 Distribution of air pollution (severely polluted) (AQI > 300) days in Lanzhou in 2003–2012

	2003	2004	2005	2006	2007	2008	2009	2010	2011	2012
January	2	–	1	6	–	1	–	–	1	–
February	1	1	–	2	–	1	–	–	2	1
March	2	2	–	8	2	1	3	5	3	4
April	3	–	2	5	1	–	2	2	5	5
May	–	–	–	–	–	2	–	–	–	1
June	–	–	–	–	–	–	–	–	–	1
July	–	–	–	–	–	–	–	–	1	–
August	–	–	–	–	–	–	–	2	–	–
September	–	–	–	–	–	–	–	–	–	–
October	–	–	–	–	–	–	–	–	–	–
November	1	3	3	–	–	–	–	3	1	2
December	2	1	7	8	1	–	–	2	2	–

The above analysis revealed that Lanzhou was a serious polluted Chinese city, so it was necessary to study the air pollution characteristics of Lanzhou in the last 10 years urgently. Table 4.1 shows the environmental air quality days distribution in 2003–2012 in Lanzhou City. Table 4.1 shows that the weather number basically shows a decreasing trend (a decreasing trend was in 2006 due to the occurrence of dust weather, and it resulted in more heavy pollution days or more); a good weather number increases significantly. Overall air quality in Lanzhou was improving, but Lanzhou had still become a relatively polluted city comparing with other Chinese cities.

If the air quality is greater than or equal to a level 4 air pollution that is moderate to high, it has a great hazard to people's work, life, and health; it is a matter of special concern to people. Basing on the air pollution index (API), the value of ambient air quality in Lanzhou City in 2003–2012 was statistically analyzed. Table 4.2 showed atmospheric ambient air quality in the specific distribution of the number of days of heavy pollution in the Lanzhou City in period 2003–2012. The table showed

Table 4.3 Distribution of days of primary atmospheric pollutants in Lanzhou in 2003–2012

	2003	2004	2005	2006	2007	2008	2009	2010	2011	2012
PM_{10}	347	349	341	352	324	320	344	319	321	329
SO_2	8	14	9	7	14	31	11	8	9	12
NO_2	0	0	0	0	0	2	0	0	1	1

that in the last 10 years in Lanzhou, the level of air pollution (severely polluted) mainly occurred in winter months (in December, in January, and in February) and spring months (in March and in April), as well as a number of days occurred in autumn months (in November). So March (30 days or 25.6%), April (25 days or 21.3%), and December (23 days or 19.65%) were the most contaminated months. The 3 months accounted for more than 66.55% of all severely polluted days. In September and October, there was no heavy weather pollution for 10 years.

Comparing to the daily air quality, it was found that, all days when it was heavy pollution and when the primary air pollutants were respirable particulate matter PM_{10}, visible particulate matter was the main cause of heavy pollution that occurred in Lanzhou. In spring, the heavy pollution mainly occurred due to external sources, such as cold fronts, resulting in dust weather that occurred to the downstream of Lanzhou, carrying large amounts of dust and causing air quality to rise suddenly in Lanzhou City in a short time. The severe pollution of air quality in winter was due to the high frequency of winter inversion in Lanzhou. The thickness of the inversion layer was the highest in the winter, the wind speed was small, the static wind frequency was high, and the atmospheric stratification was stable, which inhibited the dilution and diffusion of large amounts of pollutants discharged due to coal combustion and heating. The ground pollutants were not easy to diffuse and accumulated, eventually leading to severe pollution.

The amount of primary pollutants had specific distribution in Lanzhou City in 2003–2012 shown in Table 4.3. Table 4.3 showed that in 10 years, the number of days when PM_{10} was a primary pollutant totally was 3346 days or more than 90% of all days, and the number of days when SO_2 was the primary pollutant was 123 days, while the number of days when NO_2 was the primary pollutant was only about 4 days, and the last 2 days appeared in 2008, and 1 day appeared in 2011 and 2012. The table showed that the main pollutants in Lanzhou were mainly inhalable particles, indicating that Lanzhou would face the problem of particulate pollution for a long time.

4.2 Temporal Variation in Air Pollutant Concentration

4.2.1 Annual Variation in Pollutant Concentration

Measuring the trend of environmental pollution is statistically significant. On the basis of three main pollutants (from 2003 to 2012) and dust (2003–2007), Spearman's rank correlation coefficient (r_s) of atmospheric environmental quality is used for calculation and analysis.

Table 4.4 Spearman's rank correlation coefficient in Lanzhou City

	Annual variation	Heating period	Non-heating period
PM_{10}	−0.7822	−0.7603	−0.5106
SO_2	−0.8035	−0.7836	−0.7619
NO_2	0.0268	−0.4263	0.4328
Dust	−1	−0.9	−0.8

In order to understand the trend of various pollutants more clearly, we also based on the annual changes in the heating period (winter-spring) and non-heating period (summer-autumn); the results of calculating the rank correlation coefficient were shown in Table 4.4. Table 4.4 showed that the annual changes of SO_2 decreased significantly in heating and non-heating periods. Mainly in order to improve the quality of air environment in Lanzhou and to improve the living environment, the local authority has implemented centralized heating, blue sky, and greening projects as well as other pollution control measures. Annual change of NO_2 showed nonsignificant increase trend, but in heating period, it showed significant increase trend; in non-heating period, it showed nonsignificant increase trend. Most of anthropogenic sources of NO_2 were mainly fuel combustion and vehicle flow sources.

The above analysis revealed that in Lanzhou in 2003–2012, PM_{10} and SO_2 pollution overall showed a decrease trend, indicating that the atmospheric environment had improved. In recent decades, description of SO_2 and NO_2 emissions in Lanzhou coal and petrochemical enterprises and other work had achieved great results. In achieving results, we are not optimistic; it should be recognized that its mass concentration is still at a high level, in addition to strengthen the management of motor vehicles and other mobile pollution sources. The amount of dust per unit area could be used as one of the evaluations of the degree of atmospheric pollution index; calculation of dust showed that in the whole year and the heating period, dust was significantly decreased, and in non-heating period, it decreased slowly; the calculation also resulted that the degree of air pollution in Lanzhou had been improved.

Upon the quality level of concentration of the three pollutants, an annual average concentration of respirable particulate matter PM_{10} was of 146 μg/m^3 in Lanzhou City in nearly 10 years (average range was from 129 to 193 μg/m^3) (Fig. 4.1); the maximum value of the average concentration was of 5254 μg/m^3, and it occurred in sandstorms; the minimum value of the average concentration was of 13 μg/m^3; a maximum value was a minimum value of more than 500 times, before the larger variation; the value of average concentration of PM_{10} was of 230 μg/m^3 in the heating period, and the value of average concentration of PM_{10} was of 135 μg/m^3 in the non-heating period. Mass concentration value of particulate matter in the heating period was significantly higher than the mass concentration value of particulate matter in the non-heating period, and the mass concentration values for the two periods were much more than the PM_{10} average annual air quality standard of level 2 (100 μg/m^3). The average annual concentration of SO_2 was of 61 μg/m^3 (average range was from 41 to 86 μg/m^3), the maximum value of the average mass concentration of SO_2 was of 371 μg/m^3, and the minimum value of the average

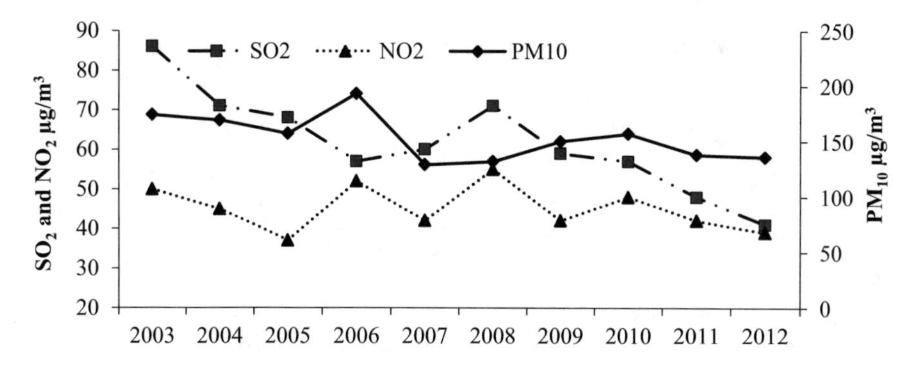

Fig. 4.1 Average pollutant concentration in Lanzhou City

mass concentration of SO_2 was of 3 μg/m^3; the average concentration of SO_2 was of 111 μg/m^3 in heating period, and the average concentration of SO_2 was of 43 μg/m^3 in non-heating period. In addition, in 5-year period of 2003–2007, the average concentration of SO_2 increased, especially in the heating period, and the mass concentration far exceeded the SO_2 air quality in annual average secondary standard (60 μg/m^3). The annual average concentration of NO_2 was of 45 μg/m^3 (average range was from 37 to 55 μg/m^3), the maximum value of average mass concentration of NO_2 was of 260 μg/m^3, and the minimum value of average mass concentration of NO_2 was of 4 μg/m^3; the average value of mass concentration of NO_2 in the heating period was of 60 μg/m^3, while the average value of its mass concentration in the non-heating period was of 37 μg/m^3.

4.2.2 Seasonal Changes in the Pollutant Concentration

Figure 4.2 illustrates average monthly pollutant concentrations in Lanzhou in 9-year period of 2003–2012. According to three main terms, the overall situation was in high winter concentration, and low concentrations were in summer and in autumn. But the annual change characteristics of pollutants were different, and SO_2 and NO_2 were single-peaked, while PM_{10} had bimodal pattern distribution.

The seasonal variation in SO_2 monthly average concentration was significant, the concentration of SO_2 in non-heating period was very low and steady, and it had the lowest level in August. Summer rainfall in Lanzhou, strong solar radiation, high temperature, strong convection, and generating atmospheric inversion layer in a short time mainly contributed to the diffusion and transport of SO_2. In winter heating period, the SO_2 concentration was significantly higher in January and in December, and the average concentration value of most of pollutants was more than 130 μg/m^3. SO_2 came mainly from emissions of coal combustion; in winter, heating coal increased the amount of coal greatly while increasing the SO_2 emissions; in

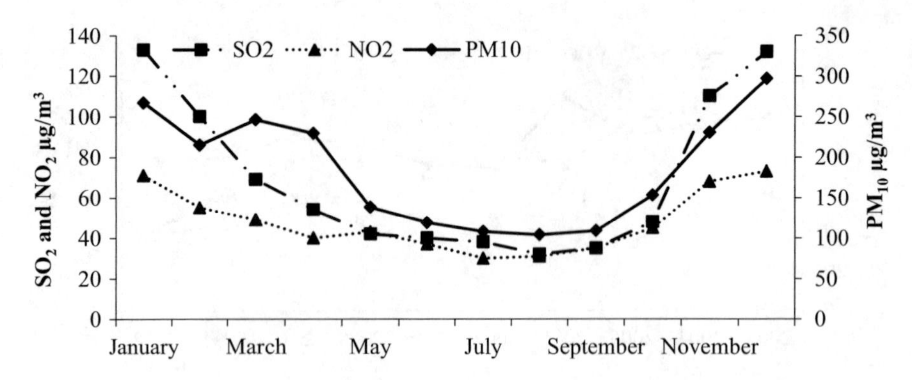

Fig. 4.2 Average monthly pollutant concentrations in Lanzhou (2003–2012)

addition to the special basin topography and meteorological conditions in Lanzhou, it was difficult to spread the pollutants, causing serious pollution of SO_2.

NO_2 is mainly an exhaust gas and industrial emission gas. It was a precursor of O_3, intense sunlight in summer, long sunshine time, and severe photochemical reaction. Concentration of NO_2 was reduced, compared with summer, the sunshine duration was short, and the intensity was weak, which was unfavorable to photochemical reaction. The monthly average concentration of NO_2 was similar to that of SO_2, which was lower in non-heating period, especially in July, and it was higher during the heating period. In a word, SO_2 and NO_2 pollutions are serious in winter, and they are the lightest in summer.

Figure 4.2 illustrated that average annual concentration of PM_{10} had bimodal pattern distribution, one of the larger values appeared in March, and the monthly average concentration might reach 240 μg/m^3; another larger value appeared in December and in January, and the monthly average concentration might reach 260 μg/m^3, with the preceding statistical heavy pollution month correspondence; the lowest mass concentration appeared in summer. This situation was mainly due to the particle emissions in Lanzhou City when the winter heating increased, and frequent inversion did not contribute to the development of atmospheric conditions, and the unique geographical environment of Lanzhou City in the wind weather; the average wind speed in winter did not exceed 1 m/s. Small wind appeared because of a stable temperature inversion in layer conditions. According to the meteorological data of Lanzhou City, in winter the inversion frequency was up to 96%, and thick inversion layer (Jiang et al. 2001) leads to very high pollutant concentrations. Another peak should lead to the dust weather effects (Wang et al. 2006), because the spring was our dusty weather-prone season. The dust transport in the upper reaches of the area, the local strong wind and dust were the main reasons for the highest peak concentration of particulate matter in Lanzhou City in spring, the frequent emergence of dust caused a sharp increase of PM_{10} concentration in a short period of time. However, during two summer seasons with diffusion conditions, good wet precipitation sedimentation, and pollution source emission reduction, particulate pollutant concentration in Lanzhou City decreased significantly.

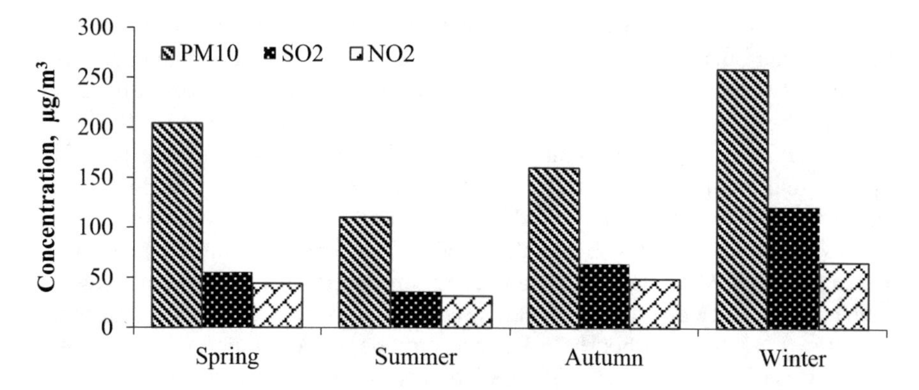

Fig. 4.3 Seasonal concentration of pollutants in Lanzhou

Seasonal characteristic of air quality in Lanzhou had obvious differences (Fig. 4.3). First, the average concentration of quarter value of SO_2 and NO_2 was the highest in winter and the lowest in summer, and the front of the pollutant concentration was changed in months in average years; according to the seasonal distribution from heavy to light, SO_2 and NO_2 concentration values were the same: winter > autumn > spring > summer. Secondly, the mass concentration value of PM_{10} varied from heavy to light in season: winter > spring > autumn > summer, while winter and spring difference was not big; spring dust weather caused increased PM_{10} concentration; however, high wind was of importance in scavenging gaseous pollutants SO_2 and NO_2; thus the concentrations of NO_2 and SO_2 were slightly lower in spring than those in autumn.

4.2.3 Daily Variations of Pollutants

According to the State Administration for Environmental Protection, for monitoring air quality in key cities of China, in 2000, two automatic monitoring stations were created in Lanzhou City. According to the data of 2016, there were five air quality monitoring stations in the city. In 2016, development of the system of atmospheric air monitoring in China involved observations over the content of fine particles in the air of 367 cities around the country (according to datacenter.mep.gov.cn); in 14 of the largest cities of Gansu Province, there were the automatic air quality monitoring stations. The original air quality "24-hour release" became "1-hour release" while retaining air quality daily data. While the previous release of another air quality index (AQI), the 24-h real-time current running within 6 h in 367 key cities, and subordinate automatic monitoring stations of three major air pollutants (SO_2, NO_2, and PM_{10}) mean hourly concentration, the public convenience was more timely and accurate understanding of the air quality situation. Five monitoring points in Lanzhou (i.e., Lanlian hotel, Worker hospital, Biological Product Institute, Railway Design Institute, Lanzhou University Yuzhong Campus) were added to the automatic air monitoring network.

The use of the "key city air quality real-time release" system obtained the data value of 5 automatic monitoring points in Lanzhou; the diurnal variation characteristics of SO_2, NO_2, and PM_{10} in spring were discussed; and the causes of the difference had been analyzed for 2 weeks since April 30, 2012, to May 14, 2012.

Diurnal changes of concentrations of PM_{10}, SO_2, and NO_2 during the periods of episode and non-episode days were studied in Lanzhou and had been illustrated in Fig. 4.4. In episode days, a day would mean the daily average concentrations of pollutants exceeding the Class II NAAQS (PM_{10} concentration was more than 150 μg/m^3, SO_2 concentration was of 150 μg/m^3, and concentration of NO_2 was of 80 μg/m^3), and in non-episode days, a day would mean the average concentration not exceeding the Class I NAAQS (24-hourly PM_{10} concentration was of 40 μg/m^3, concentration of SO_2 was of 20 μg/m^3, and concentration of NO_2 was less than 80 μg/m^3).

In Lanzhou, concentrations of PM_{10} were high, and diurnal changes were stronger during the periods of episode days. In episode days, both $PM_{2.5}$ and PM_{10} started to increase rapidly at 7–8 AM local time with an hourly rate of 15 μg/m^3 for 5–6 h, and they reached a peak concentration of ~140 μg/m^3 at noon. The concentration of the two indicators in the afternoon began to decrease slowly and at 8 PM. There was a sharp increase of PM_{10} concentrations from 100 to 125 μg/m^3, following with concentration decline until midnight. That meant that carbon particles in Lanzhou appeared from common sources of emissions, such as automobile exhaust and industrial emissions, and they were characterized by elevated concentrations of PM_{10} indicators. O_3 appeared in a high-level concentration and diurnal variation in episode and non-episode days, with a maximum concentration of 144 μg/m^3. Also, significant differences were found in the concentrations of daily changes in episode and non-episode days. A large distinction in the diurnal variation was observed in daily concentration of NO_2 and SO_2, reaching a high concentration in episode days. Increasing concentration of all the indicators occurred relatively in parallel: from 7 to 8 AM, it reached the peak concentration at 1 PM.

In addition, the concentrations of pollutants in different functional areas were also different; in 2011, annual concentration of three pollutants in Lanzhou environmental monitoring station, and in two automatic monitoring stations, and in urban area was annual average concentration of nitrogen dioxide (NO_2), sulfur dioxide (SO_2), and particulate matter (PM_{10}). Biological Product Institute, located in Chengguan District, is a representative living area; Lanlian hotel, located in the Xigu industrial zone, is a representative of the industrial zone. Chengguan District of Lanzhou City is known as a political, economic, and cultural center, as well as with developed industry and commerce; its main air pollutants are soot and sulfur dioxide, and they are produced by fuel combustion. Xigu District is known as a heavy industrial area in oil refining, petroleum chemical industry, smelting industry, and emissions of nitrogen oxides; the amount of emissions of organic compounds of ozone and olefin is larger, and it applies to the composite air pollution and to oil pollution. The two automatic monitoring stations of the concentrations of the three pollutants were compared; the value of the average annual mass concentration of SO_2 was of 49 μg/m^3 at Biological Product Institute in 2011, the average annual mass concentration of SO_2 was of 69 μg/m^3 at Lanlian hotel, in Lanlian point, at

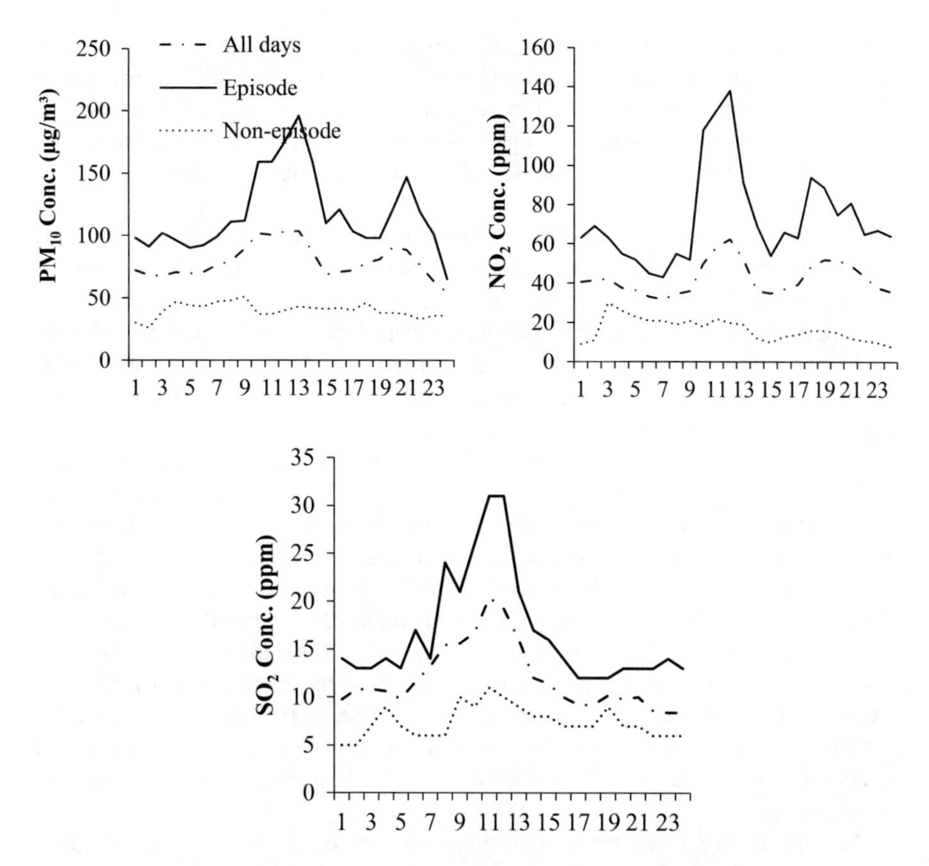

Fig. 4.4 Diurnal changes of PM_{10}, NO_2, and SO_2 in episode and non-episode days in Lanzhou

Biological Product Institute, and the average annual mass concentration of SO_2 was measured 1.4 times; values of the mass concentration of NO_2 were of 49 μg/m^3 and of 52 μg/m^3, respectively; values of mass concentration of PM_{10} were of 137 μg/m^3 and of 180 μg/m^3, in Lanlian, at the Biological Product Institute, and the concentrations were measured 1.3 times. Obviously, the Lanzhou Biological Product Institute results of monitoring of the three pollutants were significantly lower than the Xigu Lanlian hotel results, and the local impacts of industrial activities on atmospheric environment were greater than that of human activities.

4.3 Correlation Between Air Quality and Meteorological Conditions

Some studies have shown that under certain conditions of pollution sources, the level and change of concentration of pollutants mainly depend on local meteorological conditions. Local meteorological conditions could significantly affect the concentration of pollutants in Lanzhou; nonlinear correlation between concentration of pollutants and meteorological elements was usually obvious. In this paper, in 2011, three pollutants and the meteorological data of the same period in Lanzhou were selected; after the logarithm of pollution, a correlation analysis of temperature, air pressure, wind speed, horizontal visibility, relative humidity, and rainfall was conducted.

Pearson's linear correlation (*r*) was used to study the correlation between air quality and meteorological factors. The correlation coefficient from 0.000 to 0.299 was supposed as a weak correlation, the correlation coefficient from 0.300 to 0.499 was supposed as a moderate correlation, the correlation coefficient from 0.500 to 1.000 was supposed as a strong correlation, the correlation coefficient from −0.001 to −0.300 was supposed as a weak negative correlation, and the correlation coefficient from −0.301 to −0.500 was supposed as a moderate negative correlation.

Wind was an important meteorological factor affecting the dilution and diffusion of pollutants in the boundary layer; three pollutants (SO_2, NO_2, and PM_{10}) showed a weak correlation and a weak negative correlation with coefficients −0.113, −0.268, and 0.251, respectively. The correlation indicated that if the wind speed is increased, concentrations of pollutants decreased. The higher the wind speed contributed to the dilution and diffusion of pollutants in the air more, the farther the pollutant was transported in unit time, and the more it mixed with the air; the longtime breeze or static wind would inhibit the prevalence of pollutants, so that the atmospheric pollutants near the ground layer increased. Anyway, the correlation coefficient value between PM_{10} and wind was less in spring and in summer, because the wind would cause two times sand or dust weather, and PM_{10} concentration was positively correlated.

The impact of air temperature on the concentration of three major pollutants indicated a weak negative correlation with coefficients −0.296, −0.237, and −0.214 for SO_2, NO_2, and PM_{10}, respectively. Mainly due to the low temperature when coal heating and discharge, there were a lot of pollutants. In winter, in Lanzhou, it was high emission with weak convection and inversion layer, the pollutants were difficult to transfer out, and the pollution level was difficult to increase. When the temperature was high, the solar radiation was strong, the atmosphere was neutral or unstable, and the convection of the near stratum was more vigorous, which was favorable for the vertical movement of the atmosphere and the diffusion of pollutants, so that the mass concentration of pollutants was reduced.

Precipitation had removal effect on air pollutants, the correlation coefficients were of −0.36, of −0.148, and of −0.354 (correlation coefficients were tested by 0.01 significance level), the precipitation had water-soluble SO_2 and became the main part of H_2SO_3, or it was dissolved in water to form HNO_3 particles by oxidation into H_2SO_4, HNO_2, and NO_2; the raindrop impact could be attached to or dissolved in the rainwater by precipitation and ground settlement; thus precipitation was a major problem in removing air pollutants (Wei et al. 2009). The less precipitation was in winter, the higher mass concentration was of PM_{10}; summer was the rainy season in Lanzhou, and there were more precipitation and good dilution of atmospheric pollutants, reducing the mass concentration of air pollutants.

Relative humidity was negatively correlated with three pollutants; correlation coefficients were −0.363, −0.264, and −0.294, but the correlation was general. As the humidity increased, water vapor content, especially precipitation, and water vapor adsorption of the three pollutants increased, and the quality of PM_{10} improved, so that particles settled to the ground. The relative humidity was positively correlated with the concentration of three pollutants in winter, but it was negatively correlated in other seasons. Apparently, average relative humidity was low because of winter precipitation; when the relative humidity was large, the fine particles in the air were easily adsorbed by the condensate. At the same time, due to the lower temperature in winter, higher atmospheric pressure caused the atmospheric mixing height to become lower, and the boundary layer declined with the appearance of inversion temperature. The accumulation of water vapor in the lower atmosphere was not conducive to the diffusion of pollutants. In addition, the small wind speed in winter made it difficult for the pollutants to migrate horizontally, thereby increasing the concentration of pollutants.

In addition, the visibility and the concentration of three pollutants indicated a negative correlation; the correlation coefficients were −0.271, −0.219, and −0.534; the pollution was more serious, and the visibility was lower. Anyway, in spring, PM_{10} had a very good correlation with visibility, and correlations of SO_2 and NO_2 were relatively low; mainly because of dust weather caused by high concentration of PM_{10}, visibility decreased, while the wind reduced the concentration of SO_2 and NO_2; the correlation between three pollutants and visibility in summer was low, and at the same time, visibility had a negative correlation with rainfall; obviously, as summer rainfall was the main season in Lanzhou, three pollutants had a good scavenging effect, and at the same time, they caused a decrease of visibility in the city, so visibility was mainly affected by relatively high rainfall.

Chapter 5
Level of Pollutants Concentration in the Atmosphere of Lanzhou

We considered the data of gross pollutants and got an overall picture of air pollutions, largely originated from the emissions of industrial enterprises. A lot of low and irregular sources (transportation, boiler, heating, etc.) weren't taken into account. However, in our opinion, the most objective and precise criterion for the air pollution was a concentration of impurities or the mass of the pollutants in the unit volume of air (mg/m^3 or μg/m^3).

The level of air pollution increased though statistical characteristics such as the average and maximum concentrations of impurities. Air quality index (AQI), as an integrated indicator, was calculated basing on a consideration that there were more than several harmful substances in the atmosphere.

Source data for the analysis of atmospheric pollution was obtained by the input 5 air quality monitoring stations in Lanzhou City in period since 2013–2016 (Fig. 2.2).

5.1 Total Level of Pollution

In order to study the total level of pollution in the main city of Gansu Province, the integrated target of air quality index (AQI) was analyzed. The analysis revealed the following:

1. A continuous decrease of air quality index was from 2001 to 2005, and suddenly increased in 2006; in 2007 the API value was the smallest in 11 years, accounting for 90, air quality was the best, and then the API value increased year by year, by 2011 it increased to 95. According to the API average, 2007, 2008, and 2011 were the best years of air quality for the last 11 years; API average was 90–100. According to China's air quality classification criteria, air quality attained the second level of standards, and air quality was good; in the rest of the year, the annual average value of API was above 101, the air quality could not attain the national level of two, and the average value of API in the past 11 years was as

M. Filonchyk, H. Yan, *Urban Air Pollution Monitoring by Ground-Based Stations and Satellite Data*, https://doi.org/10.1007/978-3-319-78045-0_5

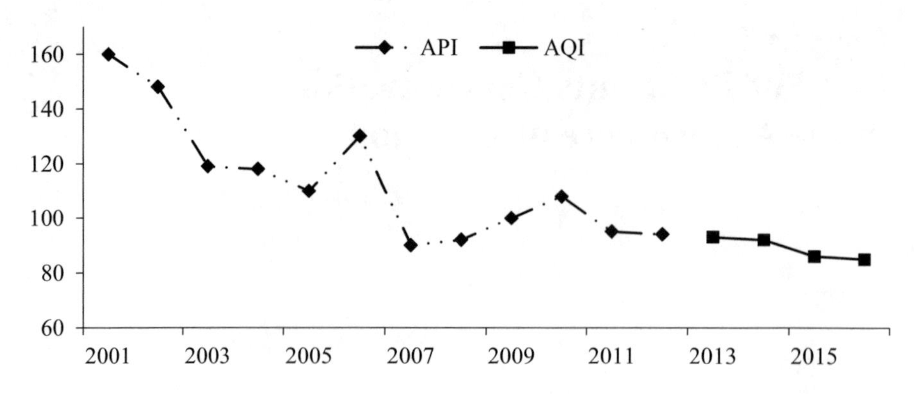

Fig. 5.1 Dynamics of air pollution index (API in 2001–2012 years and AQI in 2013–2016) in Lanzhou City

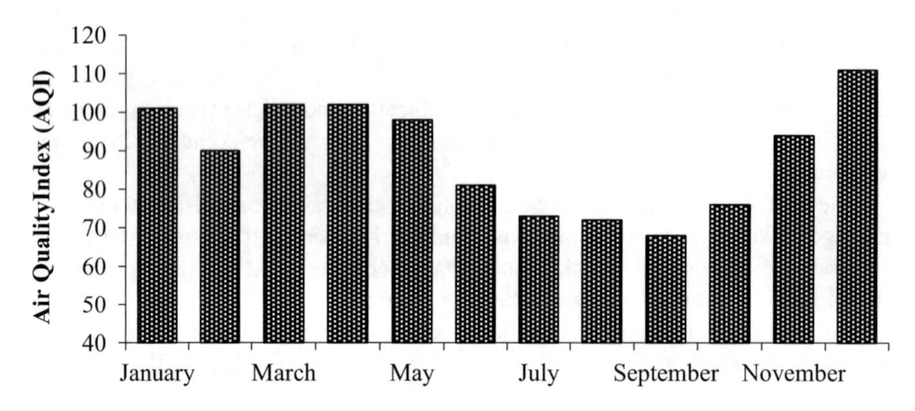

Fig. 5.2 Average annual variation in the level of air pollution for the period of 2013–2016

high as 115, reflecting the seriousness of air pollution in Lanzhou. Figure 5.1 illustrated that the air pollution was serious, the API average annual value was higher in the years, and the API standard deviation was also higher, so, in 2006, there was the most serious pollution, than in 6 years before, and in 6 years after, the API standard deviation also attained a larger value. In period of 2013–2016, the AQI was gradually decreasing and attained the lowest value in 16 years.

2. Although the level of air pollution has gradually reduced every year, it has still remained high. According to the zoning made by Seleguei and Yurchenko (1990), the natural self-cleaning capacity of the cities was not able to cope with the current load in the air. This has led to environmental disorder and to variability of elements of different environments.
3. The deterioration of air quality was observed since November to January (Fig. 5.2). It contributed to the increased load on the thermal power plants, and it was unfavorable for the dispersal of emissions in climate conditions of the terrain. Secondly, a pronounced maximum in the annual AQI cities was observed

in spring, coupled with the role of dust storms in this period. Improvement of the air quality was mainly observed since June to October. In this period of the year, general circulation factors contributing to the dispersion and leaching of contaminants from the atmosphere were of importance.

Taking into account that the total air pollution has mainly been dependent on the content of its specific impurities, further, there are studies about the distribution of concentrations of pollutants in urban area.

5.2 Statistical Characteristics of Pollutant Concentrations in the Atmosphere of Lanzhou in 2013–2016

The impurity concentration in the atmosphere (*C*) of the industrial city was determined by the influence of local or more general regional and global macroscale factors. Then.

$$C = C_{\text{loc}} + C_{\text{glo}} \tag{5.1}$$

In an urban environment, where the bulk of the pollution sources was concentrated in lower atmosphere, local pollution was dominant. This term was very unstable over time and depended on the mode of operation of enterprises, vehicles, and associated weather conditions. After its removal from sources of pollution, the impurity content of the atmosphere was approaching to regional background. It could be usually seen in the middle troposphere, where, due to macroscale sharing and self-purification processes, more or less regular distribution of impurities was determined.

5.2.1 *$PM_{2.5}$ and PM_{10} Concentrations in Lanzhou*

According to National Ambient Air Quality Standard of China (GB3096–2012), pollutants were selected into two classes of maximum allowable concentration of pollutants. Class I standards apply to special regions such as national parks (with concentrations $PM_{2.5}$ 15 μg/m^3 and PM_{10} 40 μg/m^3). Class II standards apply to all other areas, including urban and industrial areas (with concentrations $PM_{2.5}$ 35 μg/m^3, PM_{10} 70 μg/m^3, SO_2 60 μg/m^3, NO_2 40 μg/m^3, O_3 hourly 200 μg/m^3 and CO 10 mg/m^3). In Lanzhou City, for 4 years, average $PM_{2.5}$ concentrations had been reducing from 67 μg/m^3 in 2013 to 41 μg/m^3 in 2016. Concentrations of PM_{10} were also characterized by a gradual decrease of performance from 144 μg/m^3 in 2013 to 106 μg/m^3 in 2016 (Fig. 5.3). Although, the level of contamination was decreased, it had still remained high, which greatly exceeded the World Health Organization (WHO) annual average guideline value of 10 μg/m^3 (WHO 2005).

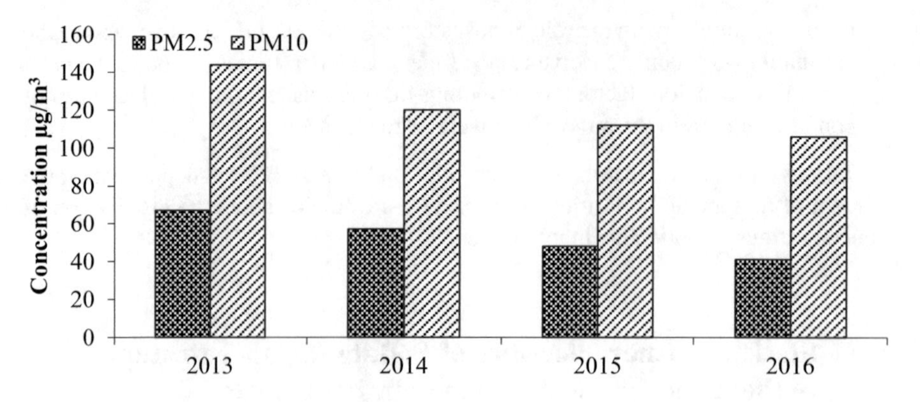

Fig. 5.3 Average concentration of $PM_{2.5}$ and PM_{10}

Table 5.1 Monitoring of mass concentration and pollution levels of PM_{10} and $PM_{2.5}$

Observation spots	City districts	Particulate matter	Daily mean range (μg/m³)	Average value (μg/m³)	Exceeding standard rate (%)
Lanlian hotel	Xigu	PM_{10}	28–894	125	66
		$PM_{2.5}$	16–312	63	80
Worker hospital	Qilihe	PM_{10}	29–977	136	81
		$PM_{2.5}$	12–256	61	74
Biological product institute	Chengguan	PM_{10}	29–838	120	60
		$PM_{2.5}$	10–485	50	42
Railway design institute		PM_{10}	30–859	113	50
		$PM_{2.5}$	12–292	54	54
Lanzhou University Yuzhong campus	Yuzhong	PM_{10}	33–651	86	14
		$PM_{2.5}$	9–185	42	20

Table 5.1 showed the five monitoring points in Lanzhou City, the different size of pollutants ($PM_{2.5}$ and PM_{10}), pollution exceeding the standard range, and the daily average range of concentrations. The monitoring time was from January 1, 2013, to December 31, 2016, with a total of 1460 days. The standard of environmental air quality was issued by the Ministry of Environmental Protection of China GB3095-2012; the quality standard was calculated as shown in Table 5.1, and within the monitoring period of 1460 days, average daily concentration of PM_{10} was at the highest monitoring point in Worker hospital. Worker hospital monitoring point located in the intersection of Lanzhou City and Xigu District was the narrowest place in Lanzhou City, and traffic congestion atmosphere diffusion capacity was weak. The highest daily concentration $PM_{2.5}$ was in Lanlian hotel. Lanlian hotel was located in Xigu District, Lanzhou City, where a large chemical factory is located throughout the country; therefore, the impact of man-made pollution sources on the this site was heavy. The average concentration of the two kinds of particles was relatively low for the monitoring point of the Yuzhong Campus, located in the eastern suburbs, less industrial, away from the source of pollution.

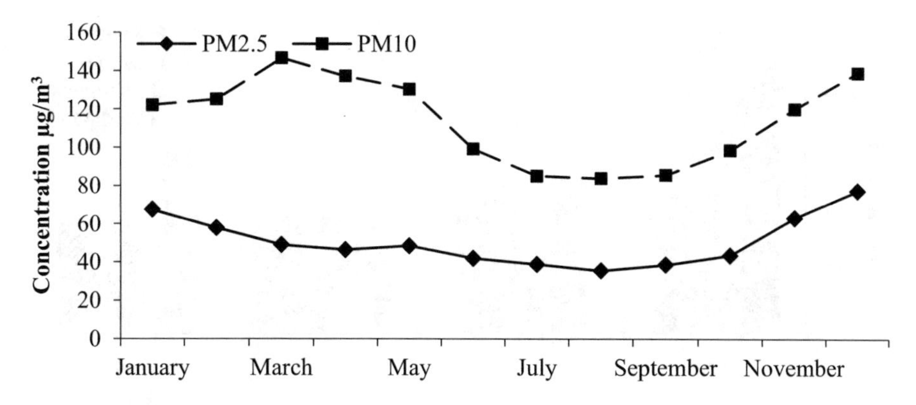

Fig. 5.4 Monthly variation in concentrations quality of PM_{10} and $PM_{2.5}$ in Lanzhou

Figure 5.4 illustrated five monitoring points, monitoring the average concentration of PM_{10} and $PM_{2.5}$ within 12 months. The mass concentration of $PM_{2.5}$ and PM_{10}, illustrated in Fig. 5.4, showed the same variation; all concentrations showed a bimodal pattern. The average mass concentration of PM_{10} was the highest in March, then it gradually decreased, the average concentration of PM_{10} was the lowest in September, and the mass concentration of $PM_{2.5}$ was the lowest in August. Two particle average concentrations reached a peak in December. This was mainly because in the spring sandstorm prone period in Lanzhou City, located in the Hexi Corridor of the southeast entrance and affected by the Tengger Desert and the Taklamakan Desert (Wang et al. 1999a, b; Tao et al. 2007a, b; Li and Cheng 2013), the dust concentration of PM_{10} increased significantly. During the heating period from February to March, the amount of coal increased, and $PM_{2.5}$ concentration of water-soluble ions and their precursors was reached. In spring and winter, when weather in Lanzhou was drier, and precipitation and gale weather process were less, it was not conducive to air diffusion and to reduce the concentrations of $PM_{2.5}$ and PM_{10}. After April, when the Hexi region gradually became a green land, the particles transported by the outside world to Lanzhou (mainly PM_{10}) were gradually decreased; at the end of the heating period, the amount of coal decreased, and the weather process of precipitation increased, and the mass concentration of $PM_{2.5}$ in the atmosphere decreased.

Seasonal average concentration data of PM_{10} and $PM_{2.5}$, obtained from the four sampling sites, is showed in Fig. 5.5. High seasonal average concentrations of PM_{10} and $PM_{2.5}$ were observed in winter in Xigu (seasonal average concentrations of $PM_{2.5}$ were of 91 μg/m^3, and seasonal average concentrations of PM_{10} were of 158 μg/m^3). Low concentrations of $PM_{2.5}$ were observed in autumn in Yuzhong (they were of 37 μg/m^3), and low concentrations of PM_{10} were also observed in autumn in Chengguan. These results could be explained by the extensive burning of coal for interior heating. Furthermore, in winter, controlled atmosphere for high pressure is often associated with subsidence inversion, resulting in the accumulation of contaminants in the atmospheric air (Cao et al. 2005; Xu et al. 2009; Wang et al. 2013a, b). Summer concentra-

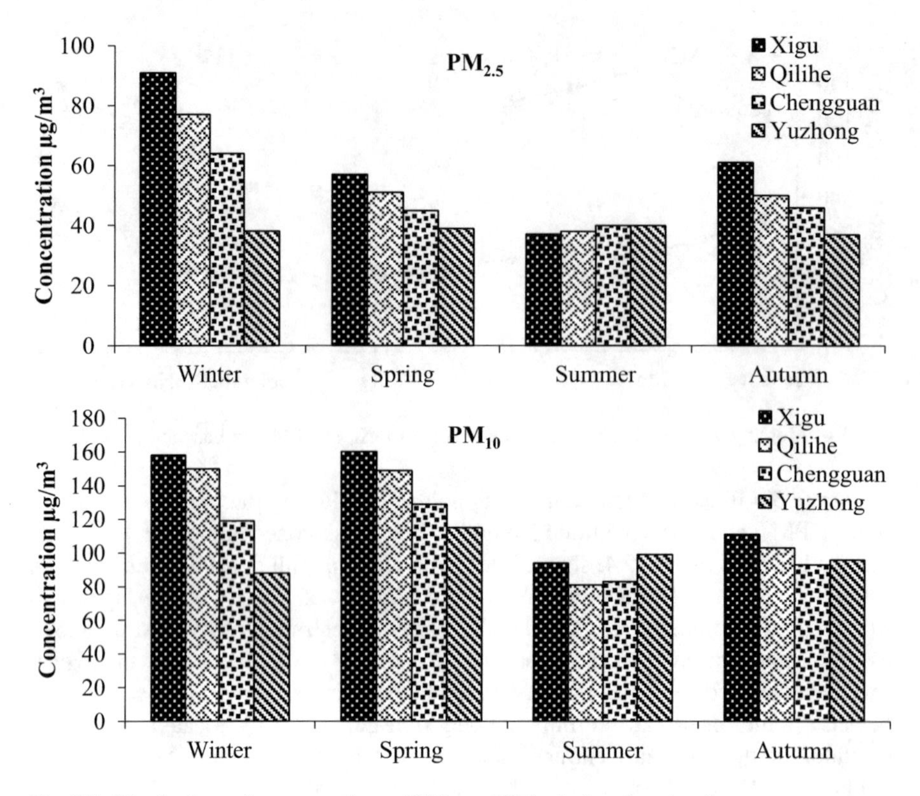

Fig. 5.5 Distributions of concentrations of PM_{10} and $PM_{2.5}$ in four functional areas

tions in Yuzhong, the level of concentration of $PM_{2.5}$, and the level of concentration of PM_{10} (level of concentration of $PM_{2.5}$ was of 40 µg/m^3, and level of concentration of PM_{10} was of 99 µg/m^3, respectively) were higher than in Xigu (level of concentration of $PM_{2.5}$ was of 37 µg/m^3, and level of concentration of PM_{10} was of 94 µg/m^3, respectively), in Chengguan (level of concentration of $PM_{2.5}$ was of 40 µg/m^3, and level of concentration of PM_{10} was of 83 µg/m^3), and in Qilihe (level of concentration of $PM_{2.5}$ was of 38 µg/m^3 and level of concentration of PM_{10} was of 81 µg/m^3, respectively), even though only sporadic and insignificant pollution sources existed in these areas. According to the Lanzhou Meteorological Administrator (LMA 2013) and other literature (Xiao et al. 2012; Zhao et al. 2013a, b), more windy weather in Lanzhou was observed in summer (in most days wind speed exceeded 2.5 m/s), indicating, that the cause for the observed values might be the predominant wind speed. High concentrations of pollutants in Yuzhong were caused by the prevailing northwest winds blowing from the nearby cities. Also, one factor was location of the Lanzhou City in the Loess Plateau as well as high wind in the summer, and dust aerosols were easily formed. In autumn, concentration of PM_{10} in the four functional areas was similar (concentration of PM_{10} was of 111 µg/m^3 in Xigu, concentration of PM_{10} was of 103 µg/m^3 in Qilihe, concentration of PM_{10} was of 93 µg/m^3 in Chengguan, and concentration of PM_{10} was of 96 µg/m^3 in Yuzhong), but concentration of $PM_{2.5}$ in Xigu (61 µg/m^3) was signifi-

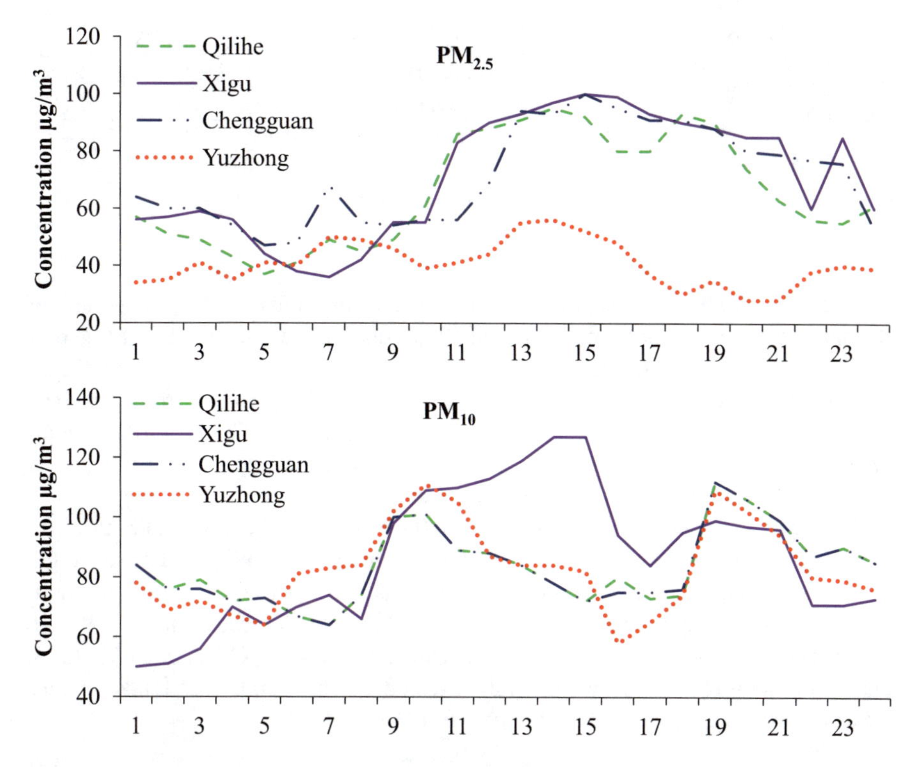

Fig. 5.6 Diurnal variations in $PM_{2.5}$ and PM_{10} in Lanzhou

cantly higher than concentration of $PM_{2.5}$ in Yuzhong (37 μg/m³) and concentration of $PM_{2.5}$ in Chengguan (46 μg/m³) and in Qilihe (50 μg/m³). In addition, a daily peak concentration of $PM_{2.5}$ was of 292 μg/m³ in Chengguan, exceeding the new national ambient air quality standard (NAAQS, 75 μg/m³) 3.8 times as much. Higher concentrations of PM_{10} were of 894 μg/m³ in Xigu, where the level of pollution was almost six times higher than NAAQS (150 μg/m³).

In the daily course of concentrations of PM, illustrated in Fig. 5.6, they were increased in the afternoon (since 1:00 p.m. to 7:00 p.m.) at the most intense mode of operation of industry and transport. At night, when many sources of emissions (vehicles, some industrial enterprises) were practically not functioning, the atmosphere was cleared of impurities. During the day, the levels of PM concentrations in the city areas had similar features, so the minimum PM emission in all areas of the city occurred in early morning about 5:00 a.m. and 7:00 a.m. Gradually we started to improve air pollution indicators with single-peaked concentrations of $PM_{2.5}$ at 4:00 p.m. with the performance of about 100 μg/m³ in all districts, except Yuzhong, where the peak concentrations were observed at 1:00 a.m. Two daily peak concentrations of PM_{10} were two peak concentrations in Qilihe, Chengguan, and in Yuzhong from 10 a.m. to 11 a.m., as well as at 8 p.m. and at 9 p.m. In the Xigu District, the reverse situation was observed, when a single peak concentrations were of 157 μg/m³at 3 p.m. and 5 p.m.

Table 5.2 Ratio of monitoring sites of $PM_{2.5}$ and PM_{10} and correlation coefficient

Observation spots	$PM_{2.5}/PM_{10}$	Ratio
Lanlian hotel	63/125	0.504
Worker hospital	61/136	0.445
Biological product institute	50/120	0.416
Railway design institute	54/113	0.477
Lanzhou University Yuzhong	42/86	0.488

The concentration values of PM_{10} included concentration of $PM_{2.5}$, so the ratio of the amount of fine particles of $PM_{2.5}$ and PM_{10} could reflect the amount of fine particles of PM_{10}. The concentrations of PM_{10} and $PM_{2.5}$ in each monitoring point in Lanzhou City were related to a numerical calculation; the ratio of correlation coefficient was shown in Table 5.2. The table shows that the ratio of $PM_{2.5}/PM_{10}$ in all areas of the city varies between 0.41 and 0.5. That indicated a great influence of the primary pollutants on the atmosphere of the city.

In winter, the coefficients R ranged from 0.43 to 0.57 indicating that both $PM_{2.5}$ and PM_{10} were the main pollutants in winter, which also explained how easy the formation of secondary and primary aerosols was in the period of strong pollution. In spring, the main sources of pollution were PM_{10}; as in all areas of the city, R have not exceed 0.4, since in Xigu $R = 0.35$, in Qilihe $R = 0.34$, in Chengguan $R = 0.34$, and in Yuzhong $R = 0.33$. This analysis revealed that atmosphere of Lanzhou City was supported by the effect of both types of PM pollutants.

5.2.2 *Concentration of Carbon Monoxide (CO) in Lanzhou*

Studies of CO content of atmospheric air are very relevant due to the constant growth of anthropogenic emissions of this toxic gas. The amount of anthropogenic emissions is globally estimated to be 200–400 million tons in year (Tanner and Law 2002). Natural sources of CO in the atmosphere are still poorly understood.

In Lanzhou airspace, carbon monoxide is one of the most common gaseous impurities. The main source of CO is vehicles, power plants, large and small boilers, and furnaces brownies. The residence time of the carbon monoxide in the atmosphere was estimated to be about 4 months. A concentration of impurity of carbon monoxide of more than 1 mg/m^3 was harmful to human's health, and to vegetation, and it is practically harmless to air concentrations. Maximum permissible concentrations are maximum onetime concentration is equal to 10 mg/m^3, and the average daily concentration is equal to 4 mg/m^3.

The average concentration of CO in the atmosphere of Lanzhou is distributed as follows:

1. In Lanzhou, level of pollution of the atmosphere with carbon monoxide was of 0.93–2.26 mg/m^3. Average annual concentrations did not exceed the maximum allowable average annual concentration for this pollutant.

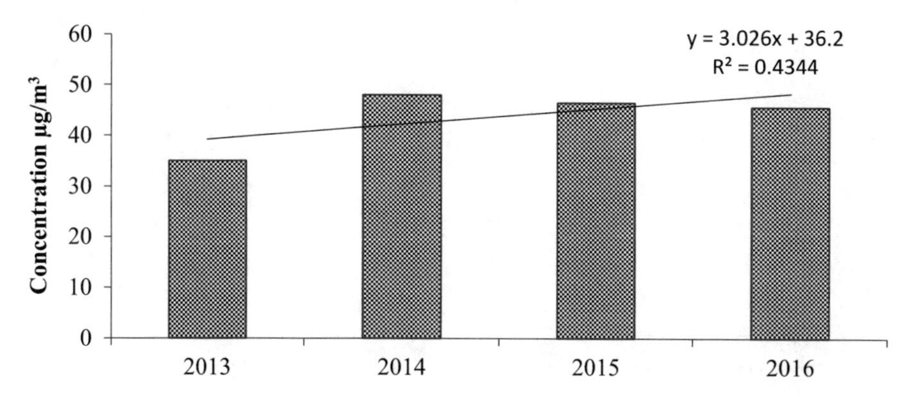

Fig. 5.7 Average concentrations of CO in Lanzhou

2. For the past 4 years, urban air pollution gradually decreased from 1.53 mg/m^3 in 2013 to 1.26 mg/m^3 in 2016 (Fig. 5.7). The linear trend equation for the city is $y = -0.085x + 1.605$, where $R^2 = 0.9691$.
3. In central areas of the city, small boilers and vehicles are usually utilized; therefore the local maxima of average concentrations of CO are registered. At increasing distance from the center to the periphery, the average concentration of CO decreased in the cities.

The annual course of the concentration of CO smoothed amplitudes did not exceed 1–2 mg/m^3 that could be explained by almost evenly vehicle operation over the year. Vehicle operation is one of the main sources of CO emissions into environment (see Fig. 5.8).

From October to March, along with the contribution of these factors, the heating furnace increased (up to 50% or more), and a thermal power plant increased along with increasing average concentration of CO in the atmosphere of the city. In this regard, the maximum fumes were often observed in winter or in autumn. So Fig. 5.8 illustrated the annual variation in the concentration of CO. Figure illustrated the highest concentrations of CO from November to January, so concentration of CO was of 1.79 mg/m^3 in November, it was of 2.26 mg/m^3 in December, and it was of 2.14 mg/m^3 in January. During the summer months, concentrations of CO were increasing in the atmosphere due to increasing traffic flow, as the city was a major transportation hub of northwestern China. So, Fig. 5.9 illustrated the average concentration of CO in 2015. The figure illustrated that increased concentrations of carbon monoxide were observed in winter. On January 11, the average concentration of CO was of 4.21 mg/m^3, exceeding the NAAQS; on January 22, the average concentration of CO was of 3.69 mg/m^3; and on December 31, the average concentration of CO was of 3.29 mg/m^3. Minimum values were the same in summer. In general, the concentration of CO in Lanzhou City was not high, and it did not exceed the NAAQS.

The diurnal variation in carbon monoxide depended on the natural variations in tension and traffic; energy emissions manifested in different ways in different areas

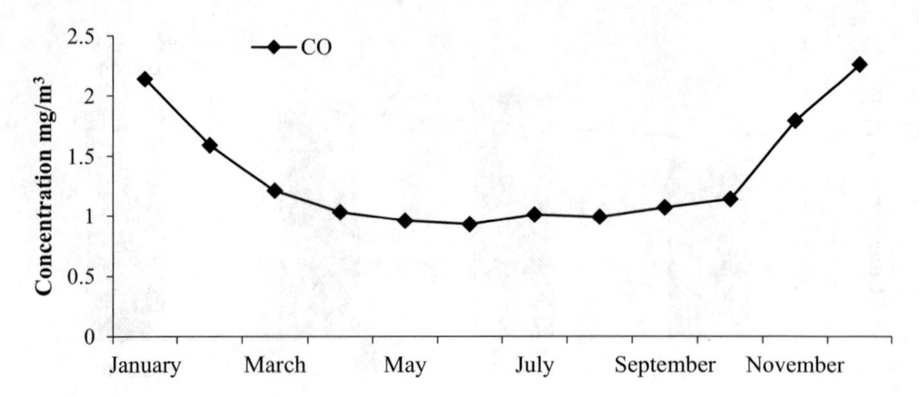

Fig. 5.8 Monthly concentrations of CO in Lanzhou

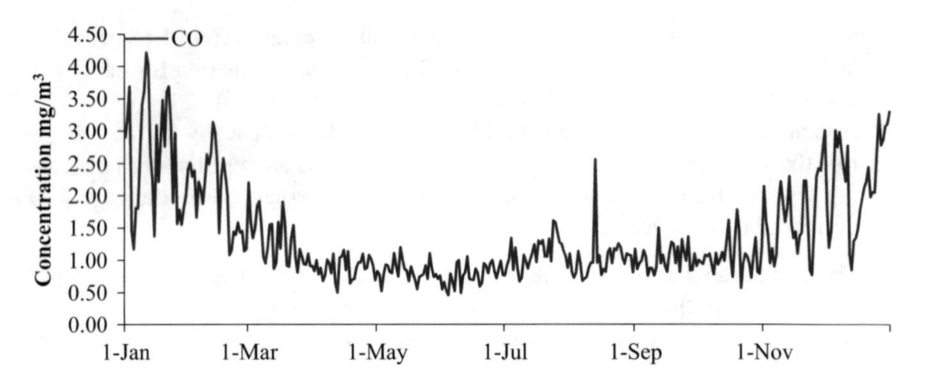

Fig. 5.9 Concentration of CO in the city in 2015

of the city. The study of values of concentrations of CO in the city showed the following:

1. The maximum concentrations of impurities coincided with the peak hour road traffic from 10 a.m. to 12 p.m. and from 7 p.m. to 8 p.m. (Fig. 5.10). The lowest gas concentration was observed at night, when traffic flow was minimal, or at 1 p.m., when the high turbulence was in the atmospheric boundary layer.
2. The impact of the private sector on the contamination of the atmosphere was indicated by the air quality monitoring stations, located at the border of the country. Thus, in Yuzhong contribution of stove heating was 50–85%, and maximum concentrations often occurred in the morning and in the evening.
3. In all areas of the city, except Yuzhong, diurnal variation in concentration of CO is clearly pronounced. In Yuzhong, the daily concentration of CO is kept to be the same; it indicated that the Yuzhong district was located outside the industrial center and it was located in the rural areas, where transport streams were not as high as in the city itself.

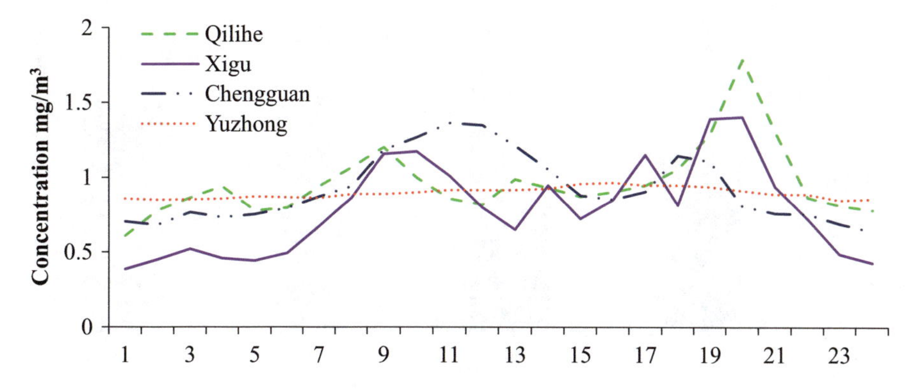

Fig. 5.10 Diurnal variation in concentration of CO in districts of Lanzhou

5.2.3 *Concentration of Nitrogen Dioxide (NO_2) in Lanzhou*

Nitrogen dioxides are emitted into the atmosphere from both natural (5×10^8 t/y) and man-made sources, as a result of the combustion process and from work vehicles (5×10^7 t/y). In cities, the values of concentration of this gas are 10–100 times as much than background values (10 μg/m^3). At a concentration of 85 μg/m^3, the gas was a serious hazard to human and plant communities. The average daily maximum allowable concentration was of 80 μg/m^3. Despite the short gas atmospheric lifetime (3–4 days), it was very prone to photochemical reactions as well as to interaction with hydrocarbons; amines formed secondary air pollutants such as aldehydes, peroxyacetyl nitrate ($C_2H_3NO_5$), carbonyl compound, nitrosamines, etc. However, some of them belonged to the most dangerous carcinogenic impurities.

Annual average concentrations of NO_2 in Lanzhou City exceed the maximum allowable concentration (40 μg/m^3) despite annual average concentrations of NO_2 in 2013. In 2013, the annual average concentrations of nitrogen dioxide were of 35 μg/m^3, corresponding to normal NAAQS. In 2014–2016, the concentration exceeded the permissible levels, so the concentration of NO_2 in 2014 was of 48 μg/m^3, the concentration of NO_2 in 2015 was of 46 μg/m^3, and the concentration of NO_2 in 2016 was of 45.6 μg/m^3 (Fig. 5.11). Trend analysis revealed a general downward trend in nitrogen dioxide pollution in the city. Trends in concentrations of NO_2 were as follows $y = 3.026x + 36.2$ at $R^2 = 0.4344$.

Many sources of NO_2 emissions were often distributed across the town square. The result revealed that the average impurity concentration distributed almost uniformly throughout the city. Maximum single concentration is often higher in areas located near highways or in the direction of the prevailing wind flow from the power station.

One maximum concentration in winter and one minimum concentration in summer were characterized by annual variations in concentrations of NO_2. Maximum concentrations in winter were on average 1.5 times as much than in summer. Therefore, the maximum concentration was observed in December, it was of 59.62

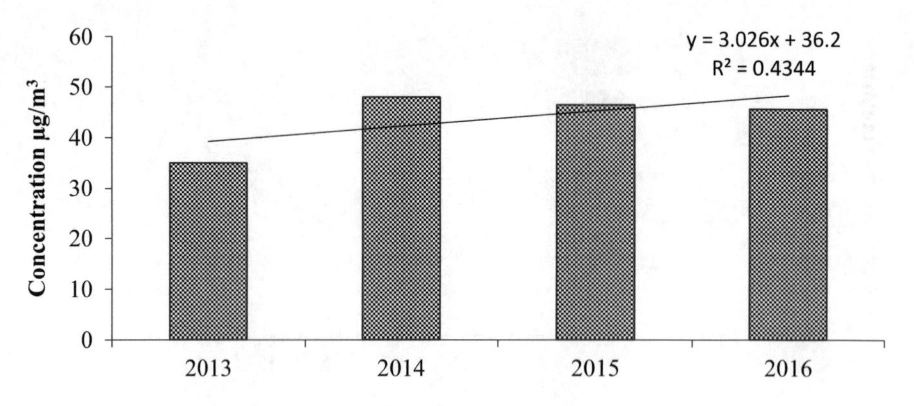

Fig. 5.11 Annual average concentrations of NO_2 in Lanzhou

Table 5.3 Seasonal concentrations of major pollutants in Lanzhou City

	Average seasonal concentrations											
	Winter			Spring			Summer			Autumn		
	Min	Max	Ave	Min	Max	Ave	Min	Max	Ave	Min	Max	Ave
$PM_{2.5}$ µg/m³	12	292	67.2	19	271	48.6	16	88	38.8	12	127	47
PM_{10} µg/m³	34	541	128.8	29	977	140.5	30	247	89.6	22	257	99.4
SO_2 µg/m³	9	75	33.7	3	53	18.1	3	36	11.4	4	40	17.2
NO_2 µg/m³	12	102	49.8	14	87	43.8	1	77	37.7	11	114	44.4
CO mg/m³	0.56	4.24	1.98	0.47	2.23	1.07	0.46	2.56	0.98	0.52	3.08	1.29
O_3 µg/m³	12	73	35.4	18	140	59.4	21	115	60.4	7	95	40.4

µg/m³, and the minimum concentration was of 35.37 µg/m³ in August (Table 5.3 and Fig. 5.12). This was consistent with the seasonal nature of emissions of NO_2 from major pollution sources.

The diurnal variation in concentrations of NO_2 was similar in all investigated areas of the city, except Yuzhong which was a suburban area, where the minimum average daily fluctuation concentrations of NO_2 were observed and where the peak concentration was at night, so concentration of NO_2 was of 45 µg/m³ at the midnight, and minimum concentration of NO_2 was of 8–9 µg/m³ from 2 p.m. to 5 p.m. (Fig. 5.13). In Xigu, in Qilihe, and in Chengguan, similar diurnal variations with several distinct peaks were observed, from 7 to 11 a.m., NO_2 indicators increased from 5 to 83 µg/m³ in Xigu. From 8 to 10 p.m., the concentration of NO_2 in three districts of the city exceeded the maximum permissible rate of NAAQS, so the concentration of NO_2 was of 58 µg/m³ in Qilihe, concentration of NO_2 was of 67 µg/m³ in Xigu, and concentration of NO_2 was of 94 µg/m³ in Chengguan.

In general, during the entire study period, Lanzhou City was supported by nitrogen dioxide pollution in excess of the National Ambient Air Quality Standard.

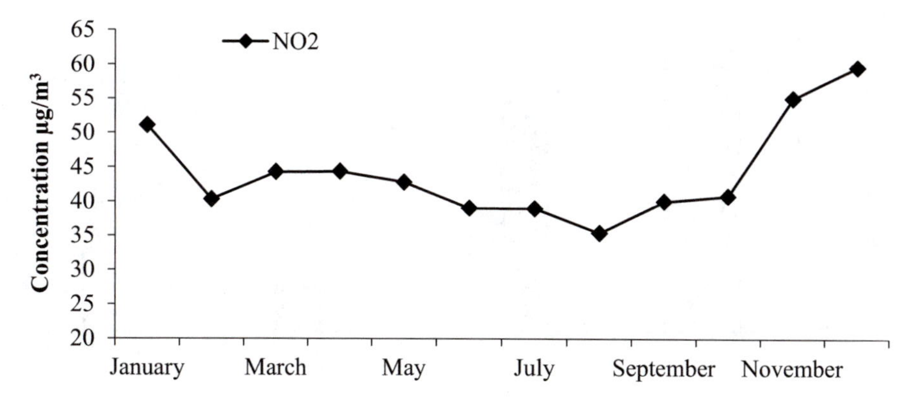

Fig. 5.12 Monthly concentrations of NO_2 in Lanzhou City

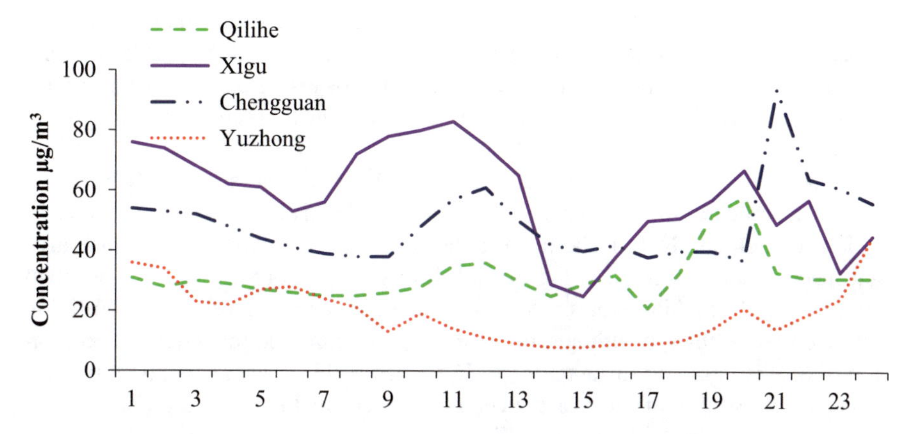

Fig. 5.13 Diurnal variation in concentration of NO_2 in the areas of Lanzhou

5.2.4 Concentration of Sulfur Dioxide (SO_2) in Lanzhou

Worldwide annual anthropogenic sources (burning fuel, etc.) emit about 100 million ton of sulfur dioxide. The contribution of natural sources is several ton of percent of total emissions. Average lifetime of sulfur dioxide over land areas is about 10 h, and average lifetime of sulfur dioxide in the heavily polluted atmosphere is about an hour. For this reason, the distance can be reduced, which can impact on SO_2 and increase its concentration in industrial centers. As a result of chemical reactions, SO_2 is converted into sulfuric acid and sulfate. Ammonium sulfate is formed very quickly, and if air humidity increases, the amount of sulfuric acid increases. According to the NAAQS Grade II standards, annual permissible concentration of SO_2 in the air is of 60 µg/m^3, and diurnal permissible concentration of SO_2 in the air is of 150 µg/m^3.

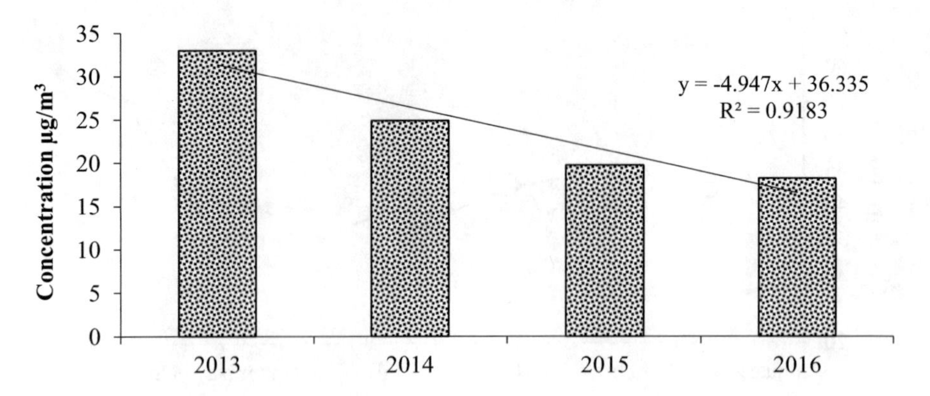

Fig. 5.14 Average concentration of SO_2 in Lanzhou

The average concentration of SO_2 in the period from 2013 to 2016 did not exceed maximum allowable concentration. The content of sulfur dioxide is time-dependent, which is illustrated in Fig. 5.14. However, trend analysis revealed a tendency to reduce the SO_2 concentrations in the city, and the linear trend equation is $y = -4.947x + 36.335$ where $R^2 = 0.9183$.

Annual variations of SO_2 were almost similar to annual variations of NO_2. At the same time, moderate and high correlation ($r = 0.576$) can be traced between concentrations of SO_2 and NO_2 in the town, indicating that their overall emission source and mechanism of formation are of high levels of pollution. In winter, the main sources of SO_2 leaching emissions as well as the maximum level of SO_2 pollution are much more, than in summer, respectively; therefore, the maximum concentration of SO_2 is observed in January (39.26 µg/m^3) and in December (36.08 µg/m^3), respectively. Minimum rates of concentration were in July (10.23 µg/m^3) and in August (11.19 µg/m^3), respectively (Fig. 5.15).

In winter, impurity removal mechanism dominated by absorbing underlying surface. In summer, there was a synthesis of sulfate ions and reduction of sulfur dioxide in the atmosphere. There was more sulfate content than sulfur dioxide gas content during the summer months; it was an indicator of the prevalence of photochemical transformations of the latter.

The diurnal variation in concentrations of SO_2 in the atmosphere of the city was a subject to slight variations during the day, compared with concentrations of CO and of NO_2. In addition, it was not the same in different districts of the city (Fig. 5.16). The most contaminated district of the city was the Xigu District, where a peak value was of 48 µg/m^3 at 12 a.m.; the other high concentrations throughout the day were observed from 8 to 9 p.m. in all areas of the city, except for Qilihe District. That could be explained by the fact that SO_2 emissions were from both low and high sources and CO and NO_2 emissions were mainly from low sources. Therefore, SO_2 dispersion in the atmosphere depended on more factors, including weather.

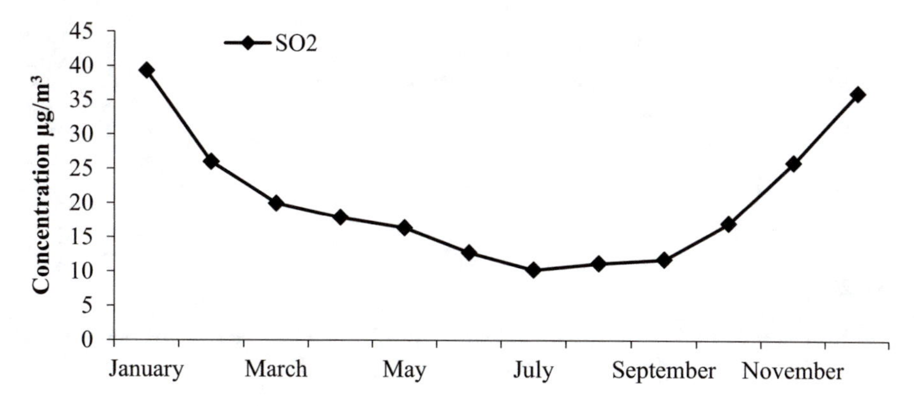

Fig. 5.15 Monthly concentrations of SO_2 in Lanzhou

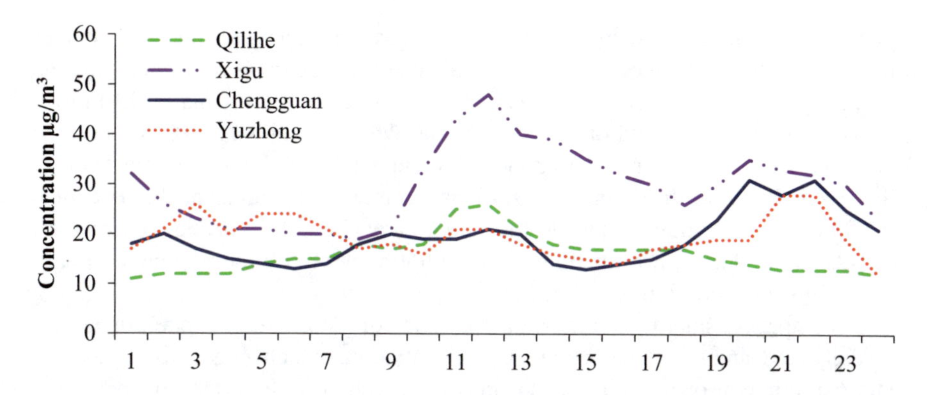

Fig. 5.16 Diurnal variation in concentrations of SO_2 in areas of Lanzhou City

5.2.5 *Concentration of Ozone (O_3) in Lanzhou*

Compared to stratospheric ozone, which protects life forms on Earth from the damaging effects of solar shortwave ultraviolet radiation, ground-level ozone is a pollutant, because there are no other air pollutants, adversely affecting human and animal health and having a negative impact on forests and crops. At high temperature and high ozone concentrations, an increase in the negative impact of each factor on the health of the population is observed. In addition, ground-level ozone is a greenhouse gas, and it has the potential of a huge impact on the problem of global climate change.

In the surface layer of the atmosphere, where there is the main source of ground-level ozone, the chemical reactions happen between oxides of nitrogen and volatile organic compounds in sunlight. Emissions from industrial plants and thermal power

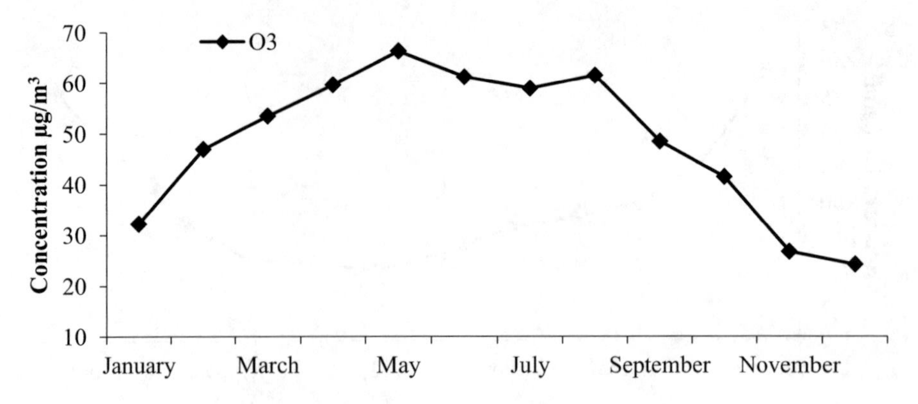

Fig. 5.17 Average monthly concentrations of O_3 in Lanzhou

plants, chemical solvents, vehicle exhaust, and gasoline vapors are the main sources of NO_x and VOC. These components are called as the precursors of ozone. They can be transported by wind for hundreds of kilometers and even in small urban resort areas, turning into a cloud of deadly gas. The World Health Organization (WHO) has classified the substances of ozone threshold actions, i.e., any concentration of this gas in the air, as the strongest hazardous carcinogens to human health. In China, the following maximum allowable concentration of Class II for ozone set according to NAAQS is diurnal 8-h maximum concentration is of 160 μg/m^3, and an hourly maximum concentration is of 200 μg/m^3.

The impact of solar radiation on the formation of ground-level ozone can be seen in daily and annual variations in its concentration, which began to increase if increasing height of the Sun reaches maximum values in the period of maximum duration of the day, and then it decreases. Values of seasonal concentrations of O_3 were high in spring and in summer; in winter, the values were minimal (Table 5.3). So the maximum concentrations of O_3 in Lanzhou City were observed from May to August, when concentrations ranged from indicators of 66.27 μg/m^3, of 61.16 μg/m^3, of 58.89 μg/m^3, and of 61.5 μg/m^3, respectively (Fig. 5.17). The minimum concentrations were of 26.7 μg/m^3, of 24.19 μg/m^3, and of 32.21 μg/m^3, respectively, from November to January. Dependence of average daily ozone concentrations on minimum relative humidity was reversed.

The diurnal variation in ozone concentration was observed as clearly visible seasonal variable. In spring, the concentration of O_3 was the highest in the afternoon at about 2~3 p.m.; in summer, concentration of O_3 was peak in the afternoon from 12 p.m. to 1 p.m.; in autumn, concentration of O_3 was peak about 2 p.m. The results revealed that peak concentration of O_3 was different in different seasons; in summer, it was the earliest one; in autumn, it was about 1.5 h later; and in spring, it was about 1.5 h later (Fig. 5.18).

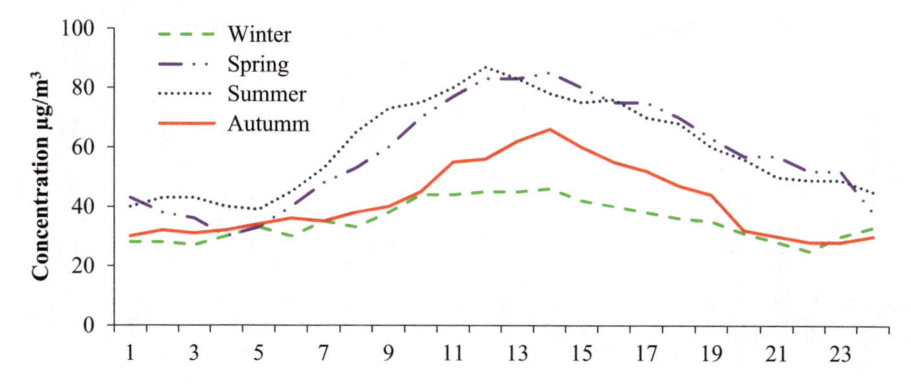

Fig. 5.18 Seasonal diurnal concentrations of O_3 in Lanzhou

5.3 Correlation Between PM and Gaseous Pollutants

The data of concentrations of $PM_{2.5}$ and of PM_{10}, used for monitoring study of districts of Lanzhou City, indicated all the 24-h but not average mean raw concentration data; that could be directly studied in different times of the concentration of particulate matter under the relevance of the analysis results which were more effective. The correlation between the particulate matter concentrations was often significant. There is often a significant correlation between the concentration of particulate matter, however, changes in pollution sources and influencing factors caused by changes in external environmental factors such as social development would inevitably lead to a corresponding change in the correlation between particulate matters. Therefore, this attracts a great interest in studying the correlation between $PM_{2.5}$ and PM_{10}, which would not just give an idea of the source of atmospheric particles, but it could also help to assess the situation of $PM_{2.5}$ pollution in the city.

Figure 5.19 illustrated the coefficient of determination (R^2) between $PM_{2.5}$ and PM_{10} data, obtained on the basis of data for the period of 2013–2016 years and showing strong correlation ($R^2 = 0.515$, $p < 0.01$). The reduced major axis (RMA) method was adopted here to calculate the regression slopes and intercepts, equations of linear trend, and the coefficient of determination, shown in Table 5.4. The average annual ratio of $PM_{2.5}/PM_{10}$ was of 0.436, indicating that the dominant coarse fraction was PM_{10} particles. That indicated a decrease in the importance of $PM_{2.5}$ in the general level of pollution of the city.

5.3.1 Correlation Between $PM_{2.5}$ and PM_{10} in Different Seasons

Figure 5.19 illustrated good correlation between $PM_{2.5}$ and PM_{10} ($R^2 = 0.515$, $p < 0.01$) in Lanzhou City. $PM_{2.5}$ and PM_{10} performed different correlations during all seasons with a coefficient of determination equal to 0.515, 0.938, 0.706, and 0.798 in winter,

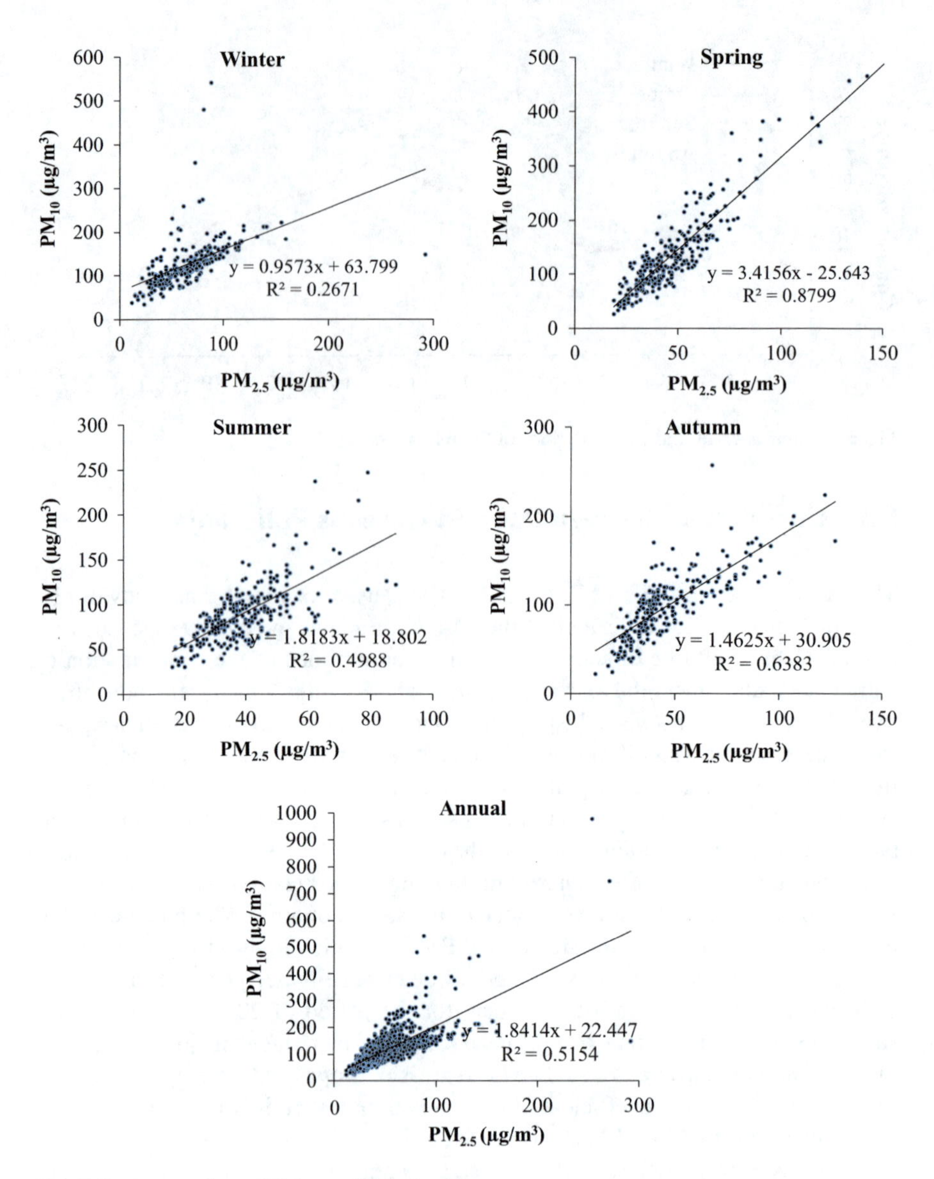

Fig. 5.19 Scatter plots of $PM_{2.5}$ and PM_{10} in different seasons

in spring, in summer, and in autumn, respectively (Table 5.4). The ratio $PM_{2.5}/PM_{10}$ in spring was the lowest, coupled with very good correlation ($R^2 = 0.879$, $p < 0.01$). It could be described so, because of the spring characteristics of dry and windy weather, which was typical for the server part of China and which released more coarse sand and soil particles into the atmosphere in spring, leading to poor quality of the atmosphere and to the minimum $PM_{2.5}/PM_{10}$ ratios. In the other seasons, the ratios of $PM_{2.5}$ and PM_{10} were comparable and indicated air pollution from both the primary and from

Table 5.4 Ratio and correlations between $PM_{2.5}$ and PM_{10}

Season		Coefficient of determination (R^2)	Pearson's correlation coefficients (r)	Ratio
Winter	$y = 0.9573x + 63.799$	$R^2 = 0.2671$	0.516	0.522
Spring	$y = 3.4156x - 25.643$	$R^2 = 0.8799$	0.938	0.347
Summer	$y = 1.8183x + 18.802$	$R^2 = 0.4988$	0.706	0.433
Autumn	$y = 1.4625x + 30.905$	$R^2 = 0.6383$	0.798	0.474
Average	$y = 1.8414x + 22.447$	$R^2 = 0.5154$	0.717	0.436

secondary sources of pollution. There was a moderate correlation between $PM_{2.5}$ and PM_{10}. $R^2 = 0.267$ in winter and $R^2 = 0.498$ in summer, respectively. Pearson's correlation coefficients (r) revealed similar correlation indices, so in spring $r = 0.938$ which played a dominant role in the total PM_{10} pollution of urban environment; increased level of PM_{10} was characterized by frequent sandstorms in the season and in the area. Winter was characterized by pollution, which included both $PM_{2.5}$ and PM_{10}. In general, in all seasons, Pearson's correlation coefficients (r) revealed a strong correlation between pollutants.

5.3.2 *Correlation Between PM and Gaseous Pollutants*

There was a similar or opposite trend in the mass concentration of atmospheric particulate matter and gaseous pollutants; this was usually due to the chemical interaction between the particulate matter and the gaseous pollutants. A large number of studies showed that sulfate, nitrate, and other organic gases precursors could be transformed into secondary particles through the evolution. So, the same method could be used to study the correlation between the mass concentration of particulate matter and gaseous pollutants and to analyze their linear correlation in four seasons. The correlation between $PM_{2.5}$ and the four gaseous pollutants was mainly studied here, because the above studies showed that the mass concentration of $PM_{2.5}$ in PM_{10} was as high as 69–87.6%. Therefore, the correlation between concentration of $PM_{2.5}$ and mass concentration of $PM_{2.5}$ represented the correlation between particles extensively. Table 5.5 showed a linear analysis of Pearson's correlation coefficient between the PM and gaseous pollutants. The correlation analysis revealed a linear positive correlation between PM and CO, NO_2, and SO_2, while there was no obvious negative correlation with O_3. In Table 5.5, the analysis revealed that a linear positive correlation existed between PM and CO, NO_2, and SO_2. The correlations (r) between PM, NO_2 and SO_2 were weak, moderate or strong for all year seasons. The following are the data in order for the winter, spring, summer, autumn respectively. The correlation between $PM_{2.5}$ and NO_2 was $r = 0.588, 0.206, 0.289, 0.642$; between PM_{10} and NO_2 was $r = 0.266, 0.153, 0.406, 0.643$), and between PM and SO_2 (between $PM_{2.5}$ and SO_2 was $r = 0.588, 0.293, 0.427, 0.71$; between PM_{10} and SO_2 was $r = 0.229, 0.251, 0.375, 0.635$. The correlation ($r$) between PM_s and CO was unstable; it meant that correlation

Table 5.5 Pearson's correlation coefficients between PM_s and gaseous pollutants in four seasons

		$PM_{2.5}$	PM_{10}
Winter	SO_2	0.588	0.229
	NO_2	0.524	0.266
	CO	0.568	0.212
	O_3	−0.427	−0.145
Spring	SO_2	0.293	0.251
	NO_2	0.206	0.153
	CO	0.275	0.191
	O_3	−0.239	−0.22
Summer	SO_2	0.427	0.375
	NO_2	0.289	0.406
	CO	0.379	0.206
	O_3	−0.436	−0.138
Autumn	SO_2	0.71	0.635
	NO_2	0.642	0.643
	CO	0.852	0.669
	O3	−0.541	−0.315
Annual	SO_2	0.605	0.323
	NO_2	0.475	0.299
	CO	0.617	0.242
	O_3	−0.44	−0.167

between $PM_{2.5}$ and CO was higher (r = 0.568, 0.275, 0.379, 0.852) than correlation between PM_{10} and CO (r = 0.212, 0.191, 0.206, 0.669), respectively, for four seasons in the Lanzhou City. This meant that the process of emissions of CO was accompanied by the emission of fine particles. Previous studies showed that emissions of NO_2, SO_2, and CO were often accompanied by the PM and that gaseous substances were associated with the formation of secondary aerosols (Tie et al. 2006). The correlation between $PM_{2.5}$ and NO_2, SO_2 and CO was the strongest during autumn, which might be caused by a lot of vehicles, cargo, that contribution was more because of the amount of exhaust gases. At the same time, the ratio of PM and gaseous pollutants and their impact on different weather conditions was different in different seasons. Correlation and different seasons showed significant difference; weak correlation was between $PM_{2.5}$ and PM_{10} and between SO_2, CO, and NO_2 in spring.

5.3.3 *Correlation Between Gaseous Pollutants*

It was generally considered that there was no simple rule to follow for the short time between the concentrations of gaseous pollutants, but it could still be through its long-term trend of mass concentration of its relevance to explore the correlation between different laws. Figure 5.20 illustrated an average annual 24-h ratio between gaseous

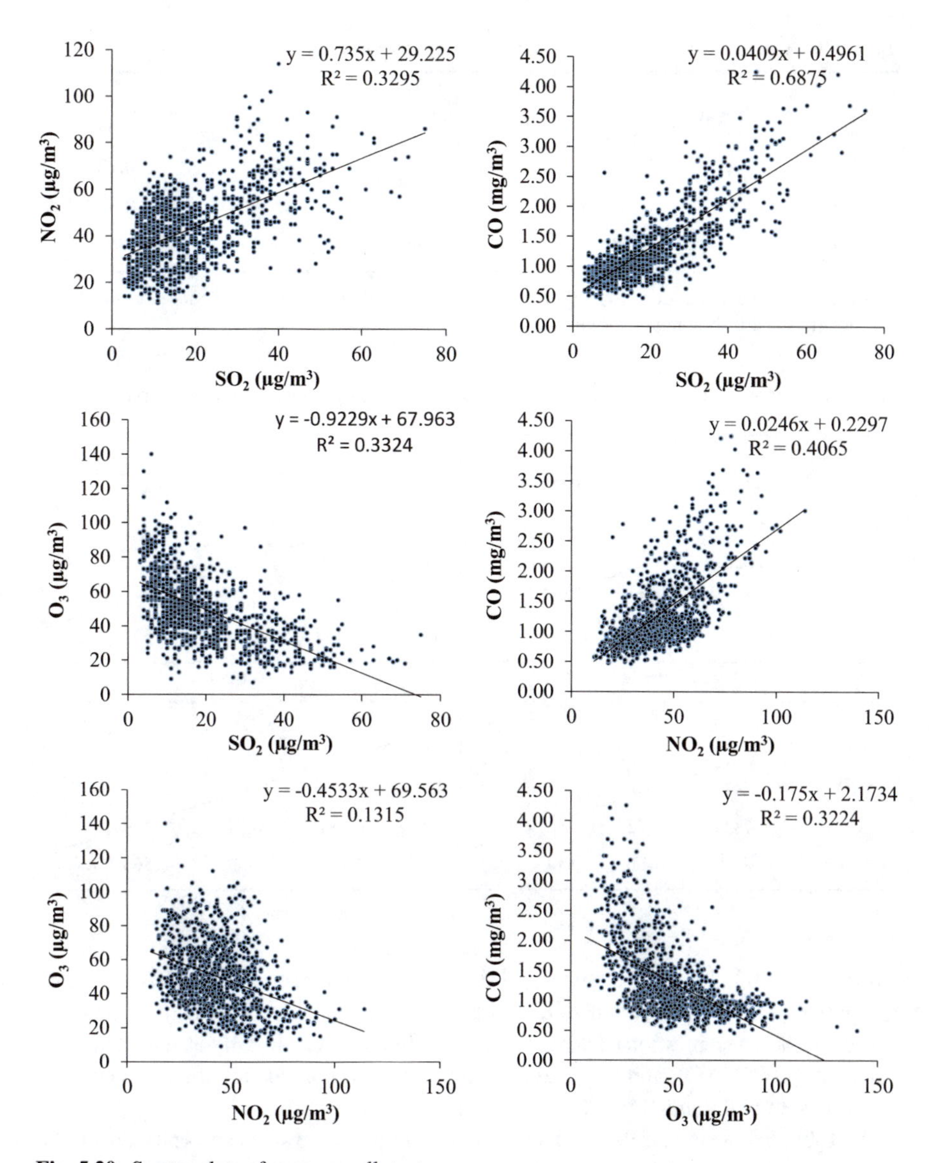

Fig. 5.20 Scatter plots of gaseous pollutants

pollutants. Analysis between gaseous pollutants, represented as coefficient of determination (R^2), and Pearson's correlation coefficients (r) were shown in Table 5.6. Figure illustrated that positive and negative correlation trends were observed between gaseous air pollutants, according to statistics of the gaseous pollutants parameter correlation, showed in Table 5.6. Coefficient of determination revealed that there were significant positive linear correlations between CO, NO_2, and SO_2 and that there was

Table 5.6 Average annual correlation coefficients of gaseous pollutants[a]

	SO_2	NO_2	CO	O_3
SO_2		0.329	0.687	0.332
NO_2	0.576		0.406	0.131
CO	0.829	0.638		0.322
O_3	-0.584	-0.36	-0.572	

[a]Coefficient of determination (R^2) values are shown above diagonal; that of Pearson's correlation coefficients (r) are shown below diagonal

Table 5.7 Pearson's correlation coefficients between gaseous pollutants in four seasons

	Winter					Spring			
	SO_2	NO_2	CO	O_3		SO_2	NO_2	CO	O_3
SO_2					SO_2				
NO_2	0.622				NO_2	0.626			
CO	0.797	0.755			CO	0.757	0.614		
O_3	-0.615	-0.561	-0.568		O_3	-0.537	-0.344	-0.521	
	Summer					Autumn			
	SO_2	NO_2	CO	O_3		SO_2	NO_2	CO	O_3
SO_2					SO_2				
NO_2	0.382				NO_2	0.529			
CO	0.483	0.322			CO	0.781	0.719		
O_3	-0.229	-0.11	-0.14		O_3	-0.620	-0.274	-0.565	

a negative correlation just between O_3. Pearson's correlation coefficients also revealed a negative correlation between CO, NO_2, and SO_2 of O_3, indicating that the impact of ozone on the concentration of other gaseous pollutants was virtually absent. The ratio between the CO, NO_2, and SO_2 had a strong correlation, indicating the close correlation between pollutants throughout the year.

In Table 5.7, seasonal Pearson's linear correlation analysis is shown between the gaseous pollutants. There was a strong linear correlation between CO, NO_2, and SO_2. There was a strong correlation between CO, NO_2, and SO_2 in the winter, in spring, and in autumn with the performance r of 0.622, 0.626, and 0.529, respectively. In the summer, there was a moderate correlation, the three kinds of pollutants followed the same evolution law, and the influencing factors might be similar, and they were influenced by the seasonal variation. According to the analysis of the impact factors of the three pollutants, the three kinds of pollutants can be preliminarily determined by coal. In summer, the impact of the two factors was relatively small, and the correlation between the concentrations of different gaseous pollutants was also affected by the seasonal variation. The correlation between O_3 and other gaseous pollutants was negative during all seasons. Correlation did not exist in summer, and it did not exist for the other three pollutants, indicating that O_3 had some mutual conversion trend with NO_2 during spring and autumn and the mechanism of evolution was more complex, for example, there was a chemical reaction between O_3 and NO_2 to generate nitrate (NO_3^-) and N_2O_5 intermediates; these substances were the main components of atmospheric secondary particles; therefore, the inter-evolution between atmospheric particles, atmospheric particulates, and gaseous pollutants was a complicated mechanism for further researching on many aspects, such as chemical reaction.

Chapter 6
Analysis of the Causes of Influencing Factors of Air Pollution in Lanzhou

6.1 Impact of Dust Storms on Air Quality

Dust weather is a kind of small probability and large hazard weather caused by the development of specific large-scale circulation background and specific weather system under specific geographical environment and underlying surface conditions, mainly in arid areas, in semi-arid areas, in desertified areas, and in farming-pastoral ecotone. Dust weather has a great destructive power, and the economic losses caused to the country and people by sand and dust weather every year are huge and incalculable (Liu et al. 2009; Qi et al. 2011). As a meteorological disaster and ecological environment problem, dust weather has already attracted the attention of scientists from all over the world. In different regions, dust weather has been studied from the aspects of synoptic, climatology, moving path, and transmission mechanism (Qian et al. 2004; Wang et al. 2005; Yin et al. 2007; Kaskaoutis et al. 2014; Dimitriou et al. 2017). Dust aerosol had a climate effect, simultaneously diffusing the incident visible light and the emergent long-wave radiation on the ground. On one hand, the dust aerosol directed scattering and absorption of radiation, resulting in direct climate effect (parasol effect). On the other hand, dust aerosol particles could change cloud microphysical processes, radiation characteristics, and precipitation, resulting in indirect climate effects (Su 2008). In addition, dust aerosols could also endanger people's physical and mental health (Chen and Yang 2001; Meng et al. 2007) and have an impact on global ecological effects (Zhang et al. 1997; Jickells et al. 2005; Huang et al. 2007; Fan 2013; Ta et al. 2013).

Dust aerosols also affect regional air quality, causing a deterioration of the air quality, affecting atmospheric visibility, and altering atmospheric composition in the city. At increasing influence of human activities on nature and at the rapid development of economy and society, the social influence caused by sandstorm weather had attracted more and more attention; the impact of the sandstorm on urban atmospheric environment had attracted a particular attention. Northwestern China is located in the hinterland of the Eurasian continent; it is one of the world's leading

M. Filonchyk, H. Yan, *Urban Air Pollution Monitoring by Ground-Based Stations and Satellite Data*, https://doi.org/10.1007/978-3-319-78045-0_6

arid areas. There are Taklamakan Desert, Gurbantunggut Desert, Badain Jaran Desert, Tengger Desert, etc., as well as large Gobi and desert lands in this region. Coupled with very little annual rainfall and high evaporation, dust aerosols result in the region's arid environment. These areas are extremely rich in sand and dust, as well as in narrow topography of the Hexi Corridor, as well as in sandstorm-prone areas of China and Asia (Wang et al. 2004; Qiang et al. 2007; Wang et al. 2013a, b). When the sand and dust weather occurs, sandstorm, dark sky, air turbidity, and poor visibility of the atmosphere, seriously affecting the people's lives, the study of dust weather in the atmospheric environment becomes particularly important.

6.1.1 Classification and Characteristics of Dust Storms

There are four major high-incidence areas of dust storms, respectively, in Central Asia, in North America, in Central Africa, and in Australia. Northwestern China is a high-incidence area of dust storms in Central Asia. In Chinese ancient books, the dust storm weather was also registered as the black wind, rain, haze, and so on. Dust and dust storms were divided into three levels by their visibility and wind. They were sandstorms, strong dust storms, and exceptionally strong dust storms.

The Concept of Sand and Dust Weather

In 1979, the China Meteorological Administration in charge with the "Specifications for surface meteorological observation" had dust, blowing sand, and sandstorm in the terminology for dust weather observation.

Sandstorm

Since 2000, when China had incorporated the sand and dust weather forecast into the business, the sandstorm had been divided into three grades: sand and dust storm, severe sand and dust storm, and extreme severe sand and dust storm. In 2006, General Administration of Quality Supervision, Inspection and Quarantine of the People's Republic of China (AQSIQ), and the Standardization Administration of the People's Republic of China (SAC) issued "Technical regulations of sand and dust storm monitoring" (GB/T20497-2006), "Grade of sand and dust storm weather" (GB/T20480-2006), and other national standards; adopted the "Specifications for surface meteorological observation," classifying the sandstorms; and defined three terms for the sand dust weather process, sandstorm weather, and sandstorm weather:

1. Sand and dust storm, sandstorm, and dust storm. High winds raised the dust of the ground, so that the air was turbid and horizontal visibility was less than 1 km within a weather phenomenon.

2. Severe sand and dust storm. Winds blew dust from the ground, making the air very turbid, and the level of visibility was less than 500 m within a weather phenomenon.
3. Extreme severe sand and dust storm. Wind blew a lot of dust from the ground, so that the air was particularly turbid, and the level of visibility was less than 50 m within a weather phenomenon.

Dust

In meteorology, the weakest form of dust is called as the dust weather. The dust existing due to remote or local produce dust or blowing sand after, formation of the dust and other fine floating air is known as the "Wild Bunch." It appears through yellow sun of distant objects. It is pale white or pale yellow and visibility is less than 10 km and more than 1 km. It is basically not clear what the wind is.

Blowing Sand

The dust, blown by the wind from the ground, makes the air quite turbid; the level of visibility is from 1 to 10 km within the weather phenomenon. Blowing sand and dust storms in local or nearby areas is caused by dust blown by the wind; its common characteristic is the visibility decreased significantly and air pollution marked in yellow color. The weather effects could appear both in cold air transit and by a cold front near distance.

Causes of Forming

The forming of sand dust weather requires three essential conditions: strong wind, strong convective instability, and a rich source of sand and dust. High winds and strong convective instability are the dynamic factors of dust and sand weather; vertical atmospheric instability will be involved in the high altitude and ground dust, with the high-altitude wind long-range transport; a rich source of dust is an essential material condition for sand dust weather. There are two kinds of the main factors affecting the formation of dust weather. They are natural and man-made factors. Natural factors such as weather and climate change are not a subject to human control. For example, some scholars have found out that the cyclic variation in the climate is the main cause of the sand-dust interannual variability, and the regional atmospheric circulation characteristics would affect the spatial distribution of dust weather; the nature of the underlying surface would affect the activities of the sand (Wang et al. 2009a, b). Han et al. (2009) studied the dust weather in Minqin County and got the influence of temperature, precipitation, and wind days on the dust weather in Minqin County. The increase of the desert area, the degradation of the grassland, the destruction of the surface layer and the vegetation cover, the expansion of desertification, the global warming and the ecological degradation are caused by the greenhouse gases emitted by human activities.

When increasing the flow of wind power and passing over the unfastened particles, the latter begin to vibrate and then “ride.” When repeated strikes on the ground happen, these particles create a fine dust, which rises in the form of suspension.

The particles are released mainly due to the dryness of the soil and high winds. Gusts of front winds may occur due to the air-cooling zone where there are thunderstorms with rain or dry cold front. After passing a dry cold front, convective instability of the troposphere can contribute to a dust storm. In the desert regions, dust and sand storms frequently occur due to lightning downdrafts and associated increasing wind speed. The vertical dimensions of the storm are determined by the stability of the atmosphere and the weight of particles. In some cases, the dust and sand storms may be limited to a relatively thin layer due to the effect of temperature inversion.

6.1.2 Regional Pollution Characteristics of the Dust Weather

After the sand dust weather started, it was strengthened by the supplement of sand and dust on the way of the sources, as the weather system moved to the downstream direction. The serious pollution events had seriously affected the people’s life and health (Fang et al. 2003; Aili and Oanh 2015). Chinese scientists have done a lot of research on the impact of air environment of city dust on weather, but most studies were focused on the impact of a dust on air quality of a single city, and the impact of dust weather on air quality in the city area of study is relatively small. Through the daily air quality data released by the Ministry of National Protection, Northwestern China and North China mainly focus on the city’s atmospheric environment quality daily data analysis, impact of dust weather on northern Chinese cities regional atmospheric environment. Figure 6.1 illustrated the evolutionary trend of air pollution index for 3 years in Northwestern China in 2014–2016 years. The figure illustrated that the air pollution index in the cities of northwestern part had obvious annual variation, in the winter heating period, heavy pollution weather would be in some cities (such as Urumqi, Xining, and Lanzhou), and this kind of pollution was mainly related to local pollution emissions and to special geographical environment.

Table 6.1 showed the severe weather pollution statistics in nine main cities in the north and northwestern parts of China. The table showed that Urumqi, Shijiazhuang, and Beijing were three cities, where there was the most heavy pollution weather; heavily polluted weather mainly occurred in Urumqi in winter; during the winter, heating period was the main cause of more pollution in coal-burning days. In Lanzhou, and Beijing, and other cities, heavily polluted weather occurred in spring; the heavy pollution weather mostly occurred in Hohhot in spring; it was found that dust weather had a great impact on Lanzhou, Beijing, and Shijiazhuang; in these three cities, the number of sand dust days, when the moderate pollution was the heaviest, was far higher than it was in some other cities.

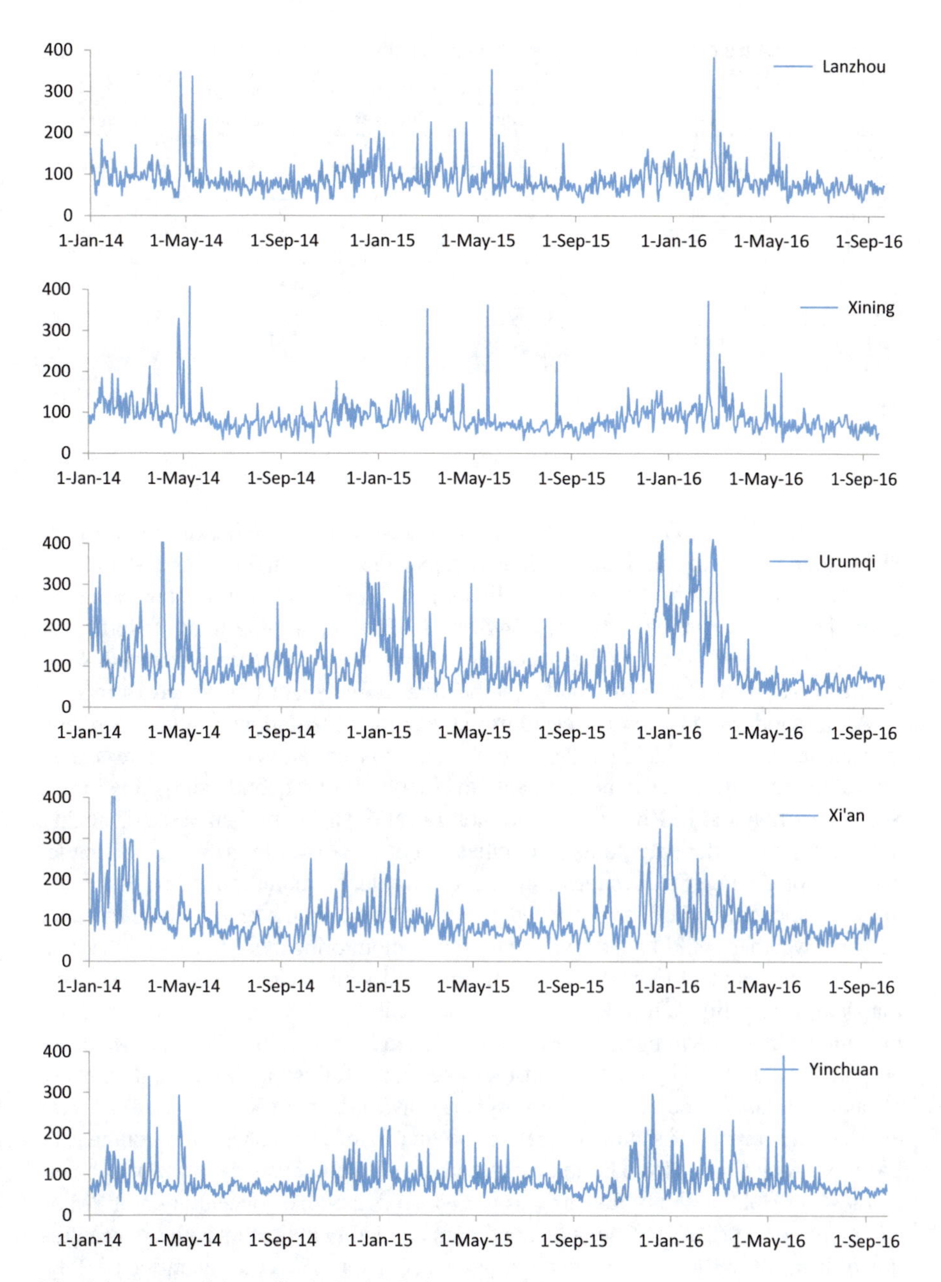

Fig. 6.1 Trends of air quality index in Northwestern China

In spring, in March and in April, there was a high incidence of dust weather in Northern China. During 4 days from April 22, 2014, a strong regional sandstorm occurred in Northwestern China due to the strong cold air moving southward from Mongolia. Dust weather covered a range of local areas in South Xinjiang basin, in

Table 6.1 Seasonal distribution of AQI ≥ 300 in major cities in North and Northwestern China

City	AQI ≥ 300	The season of the year (number of days)			
		Spring	Summer	Autumn	Winter
Lanzhou	5	3	–	–	2
Xining	5	4	–	–	2
Urumqi	34	5	–	–	29
Xi'an	7	–	–	–	7
Shijiazhuang	54	4	–	11	39
Taiyuan	1	–	–	–	1
Beijing	31	1	–	11	19
Yinchuan	2	2	–	–	–
Hohhot	1	–	–	1	–

Qinghai, in Western Gansu, in Inner Mongolia Midwest, in Ningxia, in Northern Shaanxi, in Northern and Southern Shanxi, in Northern Henan, in Central Shandong, in Eastern Sichuan basin, in Western Jiangxi, in Western Gansu, in some areas of Central and Western Inner Mongolia, in Eastern Ningxia, in some areas of Northwest Shaanxi and South Xinjiang basin sandstorms or strong sandstorms occurred, wind speeds were from 5 to 13 m/s, and in some areas, wind speeds exceeded 14 m/s.

As the sand and dust were directed from the northern part to the southern part, and the sand lead to dust creating in the air of the city, concentration of PM_{10} increased by several times. That affected the air quality in cities and consequently air quality index. So, in Lanzhou and in Xining, Air Pollution Level 6 was more than 300. Air quality index reached moderately polluted in cities, such as Xi'an, Shijiazhuang, Chengdu, and Hohhot. As Fig. 6.2 illustrated, some urban air quality pollution indices indicated that the concentration of PM_{10} reached a very high level in the cities, where the sand and dust weather could be observed; the ground monitoring system showed that this was a direct result of dust storms. From the sand source close to the city such as Lanzhou, air quality AQI index value in Xining City quickly reached 300, while the maximum value in Xining in Lanzhou was reached later than in 1 day. The two cities became to be affected by sand and dust weather in the following days, and the impact of sand and dust on the quality of atmospheric environment in Xining was more lasting than the impact of sand and dust on the quality of atmospheric environment in Lanzhou. As we move from north to south urban, air quality was getting worse, so on April 24, in Hohhot, the AQI value was more than 212 with a concentration of PM of 371 μg/m^3; on April 26, in Chengdu, maximum air quality indicators with a concentration of PM_{10} over 190 μg/m^3 were observed; and on April 27, a concentration of PM_{10} in Xi'an amounted to more than 310 μg/m^3. In the course of time, the impact of dust weather on urban air quality is also weakening. It can be seen that the impact of dust weather on quality of urban atmospheric environment has regional characteristics.

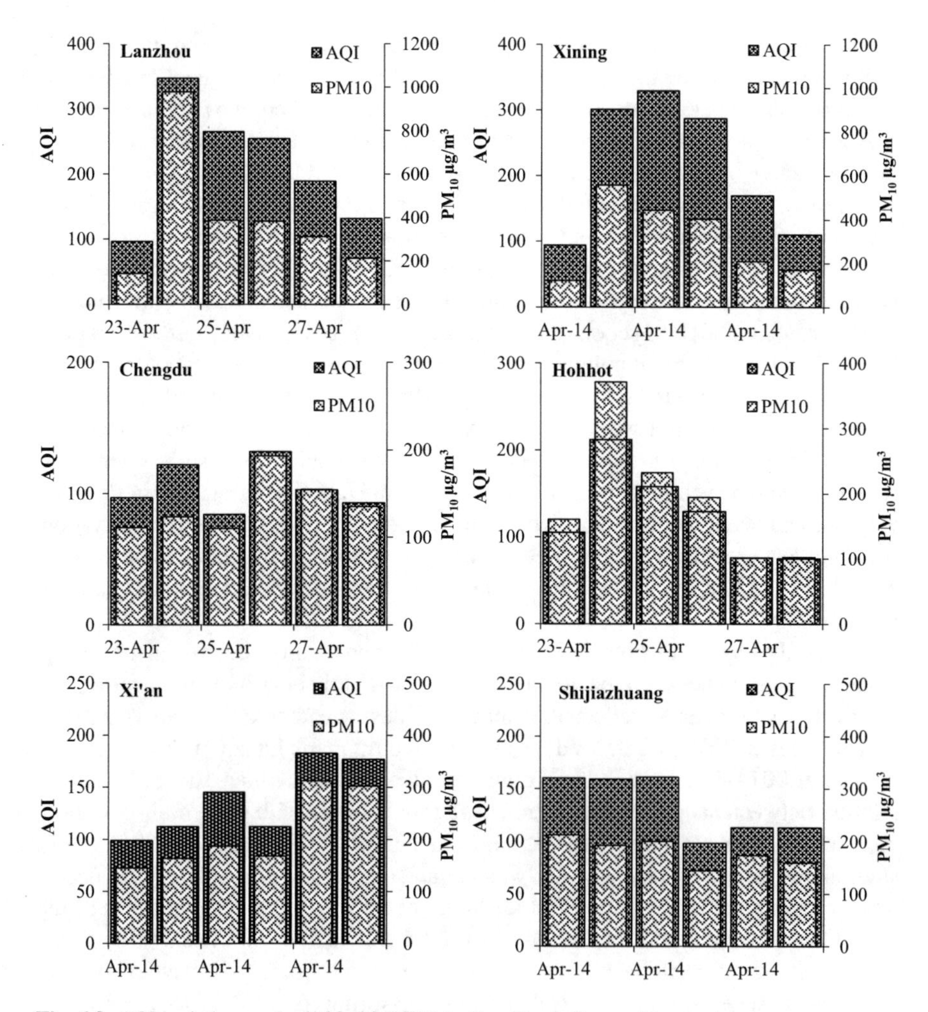

Fig. 6.2 AQI variation on April 23–28, 2016, in the cities influenced by sandstorms

6.1.3 *Effects of Sandstorms on Air Quality in Lanzhou*

Due to the special geographical conditions of Lanzhou, every year, the impacts of dust weather are more and more. Many researches have been carried out in the field of impacts of dust weather on the quality of atmospheric environment in Lanzhou. Lanzhou is located in the upper reaches of the Hexi Corridor, which includes the Gobi Desert, which has already degraded or where its steppes and croplands are degrading. At the level of statistics, it is concluded that there is a good positive correlation between the days when dust occurs in Hexi Corridor and when the concentration of particulate matter occurs in Lanzhou; when the dust storm occurs upstream, the air quality in Lanzhou would be seriously affected (Wang et al. 2013a, b). Tao et al. (2007a, b) studied and evaluated the impact of sand-dust activities on the concentration of PM_{10} in Lanzhou. Yu-Lin et al. (2006) chose to collect typical dust and non-dust weather conditions in Lanzhou City and determined their elemental composition and particle size. Zhang et al. (2010) analyzed concentration of sand dust and particle size distribution characteristics in Lanzhou in the dust storm occurred in March 2007. Filonchyk et al. (2016) investigated the intensity of pollution of PM_{10}, as well as sources of infiltration of sand and dust in the city during the summer period. A research, conducted by Liu et al. (2008), showed that in the period of 2001–2005, there were 108 days of heavily polluted events, which were related to the occurrence of sandstorms, most of which related to the strong cooling and cold weather processes in the northwest region. In addition, a small part of pollution was caused by the spring precipitation, with temperatures near the ground temperature series; after local sustained high temperatures, the main front area split weak cold air eastward and southward; in the northeastern Tibetan Plateau, warm areas strengthened the cyclonic cold low pressure system; dust weather would be generally weak in a small range of Hexi Corridor in Gansu Province area; sometimes, it would not even be up to the standard of dust weather observation, but if the low pressure system moved slowly and without precipitation, impact would extend the duration of dust weather, resulting in serious deterioration of air quality in Lanzhou City.

The study showed that the dust weather in Lanzhou mainly occurred in the spring, so the correlation between the quality of the atmospheric environment in spring and the dust weather could reflect the impact of dust weather on the air quality of Lanzhou better. Table 6.2 showed the number of days in the Lanzhou City during the springs of 2014–2016, when the air quality index was more than 200 with the correlation between sand and dust weather. Mainly coupled with data of air pollution, meteorological observation data and sand and dust weather almanac data, when considering the impact of sand and dust weather and taking into account the local occurrence of sand and dust weather in Lanzhou, but also with the upper reaches of the Hexi Corridor, sand and dust weather would occur in Lanzhou because of atmospheric environment impact. The table analyzed a total of 12–20 days of heavily polluted and severely polluted weather during the springs of 2014–2016. Description of dust weather had a significant impact on the quality of atmospheric environment in Lanzhou in spring; the air quality in Lanzhou in spring was mainly due to the dust weather, indicating, that the air quality in Lanzhou was closely related to the occurrence of dust weather. During just 4 days in winter and 3 days in autumn in the period of 2014–2016 years, air quality index had been exceeding 200, indicating the importance of both primary and secondary sources of pollution in the city.

Some relevant statistical data (Zhi et al. 2007) was based on the data of observation station of Lanzhou City in the periods of 1971–2000 and of 2001–2004, the climate characteristics of dust weather in Lanzhou and the characteristics and causes of dust weather in 2001~2004 were analyzed. The results revealed that the sand and dust weather in Lanzhou had a decreasing trend in 30 years. The dust weather in Lanzhou was based by dust and blowing sand. Dust weather was mostly observed in spring (3~5 months). Table 6.2 showed the correlation between moderate pollution weather and sand-dust weather in Lanzhou in 2014–2016 in spring period. The

Table 6.2 Seasonal changes in air quality in dust storms (AQI > 200)

Date	AQI	Air quality	PM_{10} (μg/m^3)	Visibility (km)
24.04.2014	347	Severely polluted	977	3
25.04.2014	265	Heavily polluted	388	3.7
26.04.2014	254	Heavily polluted	382	4.7
30.04.2014	244	Heavily polluted	457	4.3
09.05.2014	337	Severely polluted	746	2.6
24.05.2014	205	Heavily polluted	374	6.3
25.05.2014	232	Heavily polluted	344	5.6
28.12.2014	203	Heavily polluted	222	5.6
03.03.2015	226	Heavily polluted	360	6.4
02.04.2015	208	Heavily polluted	319	5.2
16.04.2015	225	Heavily polluted	348	6.6
18.05.2015	353	Severely polluted	466	5.4
19.02.2016	322	Severely polluted	479	6.3
20.02.2016	382	Severely polluted	541	4.6
28.02.2016	201	Heavily polluted	270	5.6
01.05.2016	201	Heavily polluted	274	6.3
11.10.2016	500	Severely polluted	949	2.8
11.11.2016	400	Severely polluted	558	3.6
11.18.2016	255	Heavily polluted	417	4

table showed that the weather with the pollution index more than 200 had a good correlation with the dust weather, and the dust weather occurred mostly. The main sand and dust weather caused heavy pollution and visibility decline in the city.

The impact of dust weather increased every year because of the special geographical conditions of Lanzhou. In this paper, the dust events in Lanzhou area were analyzed, and the impact of dust weather on atmospheric environmental quality in Lanzhou was considered. Medium visibility in dust storms varied in the range of 3–6.5 km. In a normal day, the visibility was more than 20 km. Poor visibility indicated the presence of PM_{10} in the atmosphere. So, on April 4, 2014, at a value of concentration of PM_{10} of 977 μg/m^3, visibility was just about 3 km; on May 9, 2014, at a value of concentration of PM_{10} of 746 μg/m^3, visibility was about 2.6 km. This showed the direct dependence of visibility on the concentration of PM_{10}.

6.1.4 Long-Range Transport of Dust in Lanzhou

Tracking sources and pollutant pathways in the Lanzhou trajectories were calculated using the Hybrid Single Particle Lagrangian Integrated Trajectory (HYSPLIT) model developed by Air Resources Laboratory (ARL) in the National Oceanic and Atmospheric Administration (NOAA). 72-h backward trajectories were utilized,

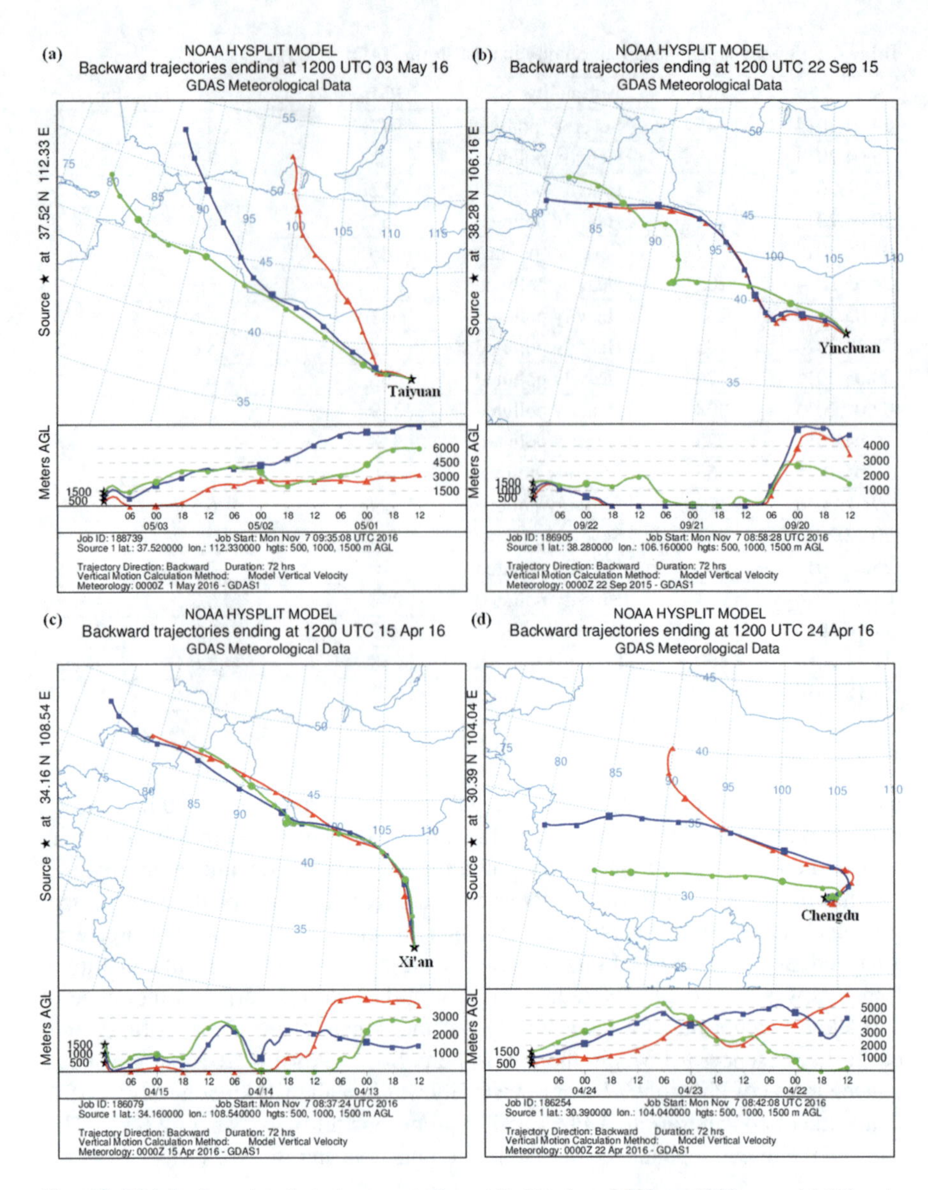

Fig. 6.3 72-h backward trajectories are starting at 3 altitudes of 500 m, 1000 m, and 1500 m in north, northwest, and southwest parts of China. (**a**) At noon on May 3, 2016, in Taiyuan; (**b**) at noon on September 22, 2015, in Yinchuan; (**c**) at noon on April 15, 2016, in Xi'an; (**d**) at noon on April 24, 2016, in Chengdu

starting at 100, 500, and 1000 m. Figure 6.3 illustrated the movement of air masses in major cities in south, north, and northwest of China. The figure illustrated that the movement of dusty air masses was mainly from the northwestern part of Hexi District; in the northern part of the country, where there were many deserts as a part

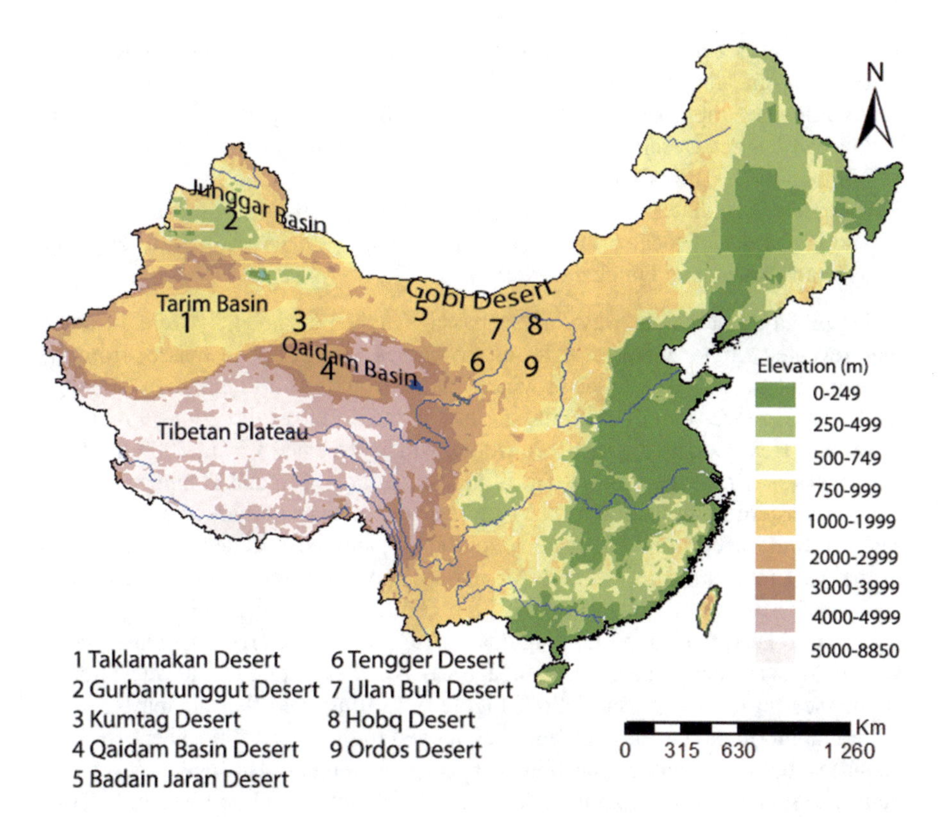

Fig. 6.4 Distributions of ten major deserts and elevation distribution in Northern and Northwestern China

of the Gobi Desert, the figure also illustrated the western direction of movement in the direction from the Tibetan Plateau (Fig. 6.4).

The occurrence and transmission of dust storms in Lanzhou are not only a serious negative impact on the economic activities of Lanzhou, but it also causes a great inconvenience for the social activities of Lanzhou citizens, especially in endangering the health of the public.

Therefore, if we could find the dust source and transmission path of dust, affecting the dust weather in Lanzhou, it would be very important to control and to improve the quality of atmospheric environment in Lanzhou. In this study, a model to calculate the sand-dust weather process in Lanzhou from 2013 to 2016 was used, as well as the path of sand-dust movement in Lanzhou was used; it was obtained from sand source and dust weather, both within the impact of dust sources and outside the sand sources.

The above analysis revealed initially the impact of sand and dust weather on quality of atmospheric environment in Lanzhou; according to the direction of the dust transmission, the main transmission path and dust source would be divided into three ways.

1. "North" is associated with air masses passing through the Gobi Desert, Republic of Mongolia, and Inner Mongolia (Fig. 6.5).
2. "Northwest" appears mainly in the northwestern part of Xinjiang (in Gurbantunggut Desert), where air masses move through Inner Mongolia and Ningxia to Lanzhou City (Fig. 6.6).
3. "Western" originates in the western deserts (in Taklamakan Desert in Xinjiang) and in the desert located in Qaidam Basin, where air masses pass through the Tibetan Plateau and Qinghai (Fig. 6.7).

Despite the fact that the ways of dust passage in the Lanzhou City were different, they revealed similar features. A lot of air masses, except for air masses from the Tibetan Plateau, moved into the province mainly from the north direction, although the air passage was mainly from the northwest and northeast directions. Figure 6.4 illustrated Hexi Corridor (Gansu Corridor), which stretched within 1000 km from the northwest (from the Central part of Gansu to Jiayuguan and Dunhuang) along the northeastern foothills of the Qinling Mountains with an altitude of over 3000 m. The width of the corridor was from 20 to 100 km, and the altitude was from 800 m in the west to 1500 m in the east. Air masses moving from the Gobi Desert and deserts of provinces bumped into the high Qinling Mountains and changed their direction to Hexi Corridor, grabbing dusty desert air characterized by existence of ground-level air masses (Figs. 6.5a, 6.6a, 6.7a). Another feature was the elevation of air masses during its transportation. Figure 6.7b illustrated that air masses were more than 4000 m aboveground and they moved to the upper atmosphere for long distances before reaching Lanzhou and lowed gradually. The height level was mainly associated with a unique landscape of Taklamakan Desert and hilly Tibet Plateau, which gradually declined from over 4000 to 1800 m as it entered Lanzhou.

So, from the above, dust storms mainly occurred in Northern and Northwestern China, indicating that this was a regional problem for the country. There were three main sources of dust entering in Lanzhou. They were Taklamakan Desert and Hexi Corridor, central Inner Mongolia Plateau, and west Inner Mongolia Plateau. Most of the dust comes from the Gobi Desert. There were two areas where there were high frequent dust events in Lanzhou; the first one was in the Western Taklamakan Desert, and another one was in the Western Gobi Desert. The dusty weather was also closely related to climate change and human activities.

6.2 Meteorological Conditions of Air Pollution in Lanzhou

6.2.1 Meteorological Conditions of Atmospheric Pollution

Accounting levels of pollutants without their correlations with meteorological parameters cannot complete the picture of the state of urban air basin. It is no coincidence that the study of this rather complicated and multifaceted communication is devoted to a very large number of works (Ipsen et al. 1969; Forsberg et al. 1993;

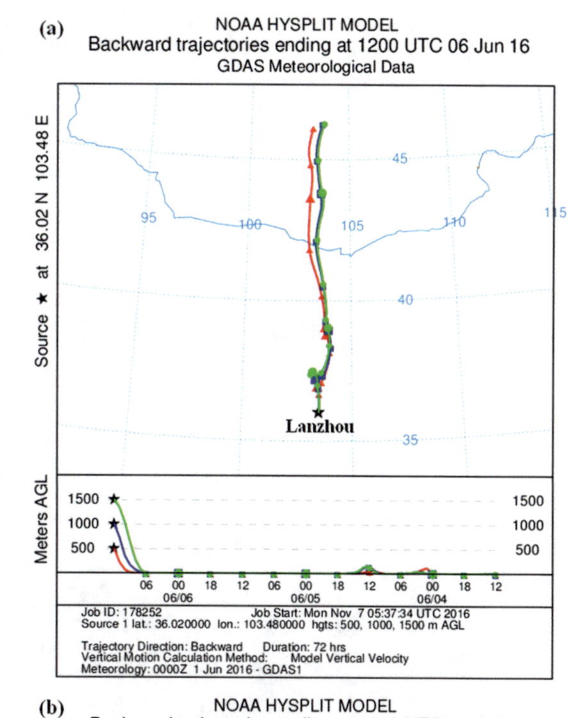

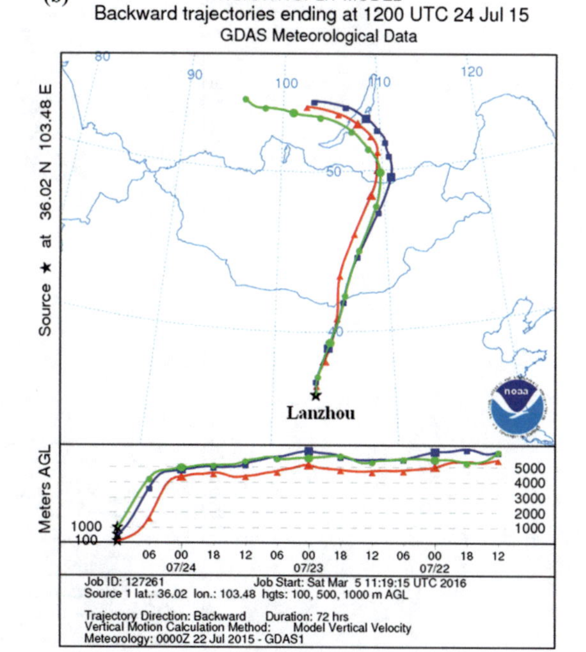

Fig. 6.5 72-h backward trajectories are starting at 3 altitudes of 100, 500, 1000 and 1500 m in Lanzhou (north direction)

Fig. 6.6 72-h backward trajectories are starting at 3 altitudes of 100, 500, and 1000 m in Lanzhou (northwest direction)

(a)

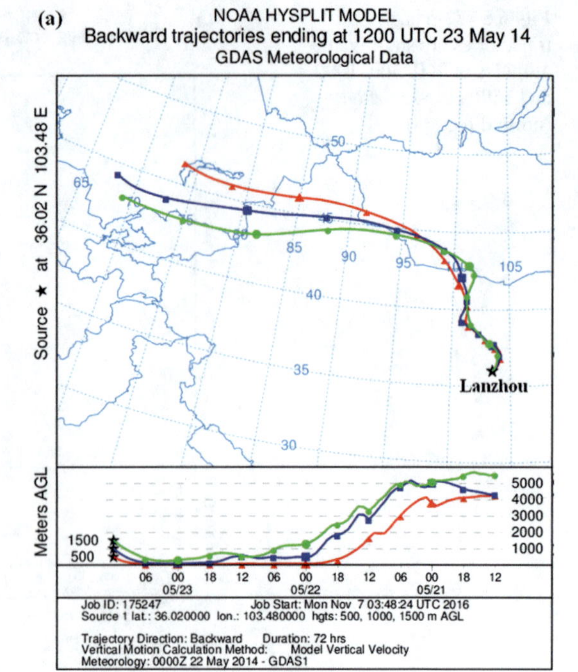

(b)

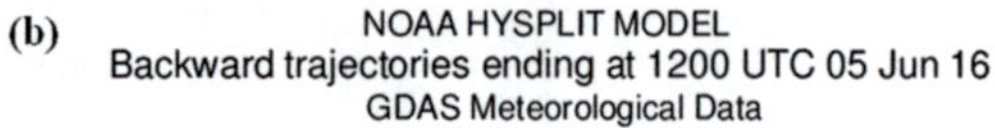

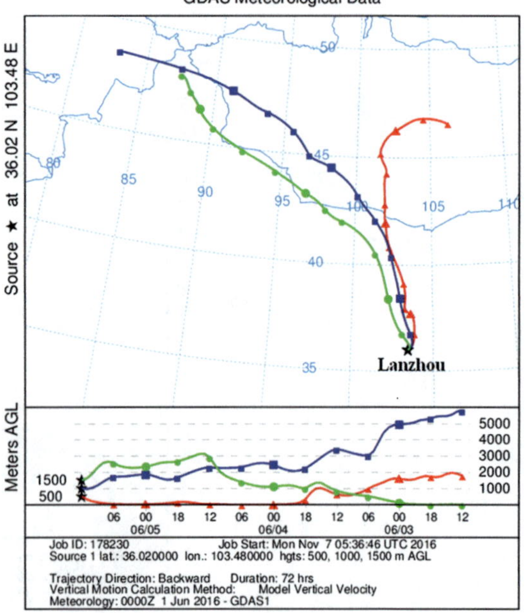

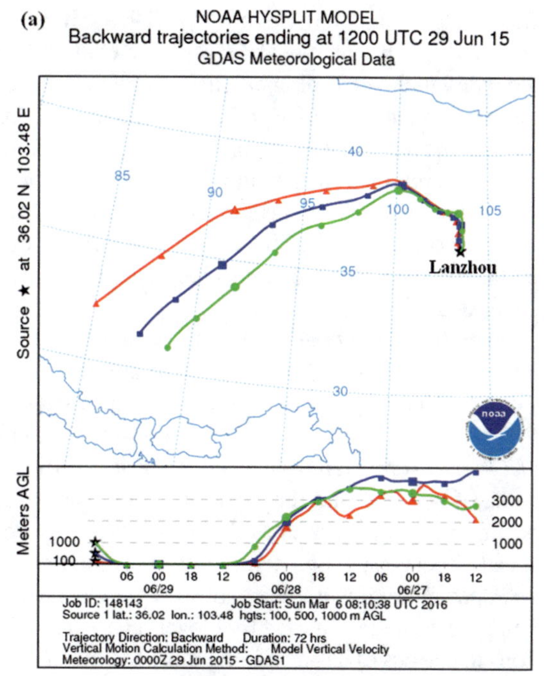

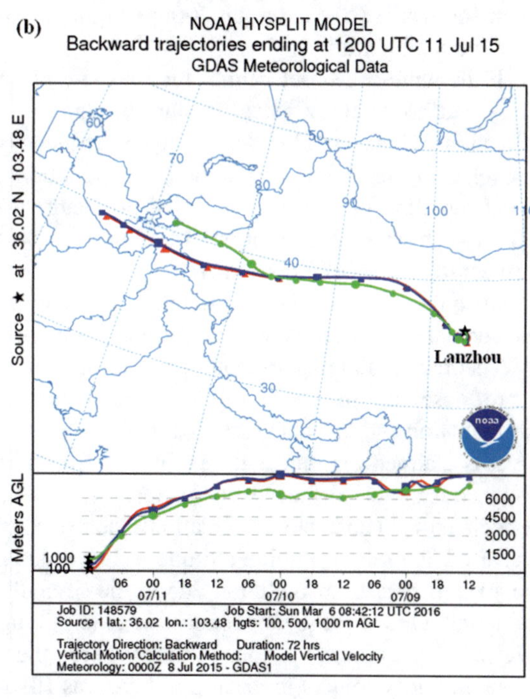

Fig. 6.7 72-h backward trajectories are starting at 3 altitudes of 100, 500, and 1000 m in Lanzhou (Western direction)

Rutllant and Garreaud 1995; Mayer 1999; Hien et al. 2002; Yang et al. 2007; Tian et al. 2014; Whiteman et al. 2014; Laña et al. 2016), focusing on the importance of the individual meteorological factors and their different influences on the formation of air pollution, as well as to identify the reasons for the accumulation of pollutants in the atmosphere and for leading to its purification.

In the opinion of many authors, the main factor determining the spread of pollutants in the atmosphere was the wind's regime (Vardoulakis et al. 2002; Kurita et al. 2010; Kuo et al. 2015). The main characteristics of the wind, significantly affecting on the turbulence of the atmosphere and the diffusion of impurities, were direction and speed of wind.

In cities, where there are many low sources of pollution growth, wind speeds are lower; they are up to 1–2 m/s. Maximum pollution by overheated torch high emissions (particularly combined heat and power (CHP) plant) was observed at a rate of 5–7 m/s. For chemical companies, considered dangerous wind speed was at a rate of 1–2 m/s, for metallurgical and refineries considered dangerous wind speed was at a rate of 2–4 m/s, and for purification highways from carbon monoxide to 70–90% occurred at a wind speed of 3–5 m/s. Thus, depending on the type of source, wind speed affects dispersion admixtures in different ways.

Especially unfavorable conditions were created when the weak winds (less than 3 m/s) remained for a long time over a large area. Repeatability of large concentrations in periods of strong winds (more than 6 m/s) was low. This conclusion is based on many works (Beckett et al. 2000; Elminir 2005), and it is associated with both low and high pollutions.

If the wind speed determines the intensity of the scattering material both vertically and horizontally when passing through the territory of pollutions, the wind direction and repeatability determine the frequency and level of pollution in particular relatively to location of a source of pollution. The most noticeable impact of wind direction on the prevalence of pollutants was from a separate source. The zones of higher concentrations of impurities were created in the leeward with respect to sources of emission areas.

In some cases, the thermal stability of the atmosphere, determining the vertical movement of air masses, was more significant to increase air pollution.

Theoretical and experimental data revealed that elevated levels of air pollution were observed in both strong turbulent exchange associated with unstable stratification and in stable stratification.

It is claimed that under the pollutions from high sources, the concentration of impurities on the surface layer was increased by enhancing the turbulent exchange associated with unstable stratification. The reason was that main impurity scattering occurred at a relatively high altitude. When pollutants were at a low level, fugitive sources increased air pollution due to the strengthening of thermal stability of air pollution. One of the main phenomena, determining the stability of the atmosphere, was a temperature inversion. In cities where there was an inversion layer above them, impurity content in the atmosphere was 10–60% higher than where there was no inversion layer (Holmes and Morawska 2006). In this case, the power and intensity of the vertical inversion layer and the height of the lower boundary of inversion and the height of the emission source were of importance.

The combination of surface inversion temperature with a wind speed of 0–1 m/s was most dangerous. Such phenomena were considered to be "stagnant," and most times they could be expected in Beaufort weather.

In general, surface and elevated inversions were unfavorable conditions for prevalence of pollutants in a certain situation. All this indicated a rather complex correlation between air pollution and the stratification of the atmosphere, which should be taken into account in the context of study of atmospheric pollution.

The impact of temperature on the level of contamination was also ambiguous. For example, in factories, thermal power consumption changed at low temperatures of air, and, consequently, the content of the products of combustion (oxides of sulfur and nitrogen) in the atmosphere was changing. On the other hand, at these temperatures, a temperature difference between the emission and the environment appeared, leading to a large rise of impurities and reducing of their impact on the ground layer. When increasing outside temperature (especially in summer), the rate of photochemical reactions increased, leading to formation of photochemical smog which was extremely dangerous to the health of the population.

If the effect of temperature and wind regime of the troposphere to the level of air pollution was the object to attract scientists' attention, then much less attention was paid to the dependence of atmospheric air pollution on other meteorological elements.

With regard to atmospheric phenomena, then mist and precipitation were most frequently mentioned (Rosenfeld et al. 2007; Guo et al. 2014; Yang and Li 2014).

The effect of purification of the atmosphere by precipitation is not in doubt what is marked by all the authors who have studied this question. In some cases, there is a direct correlation between the amount of precipitation and atmospheric pollution, and in some cases, there is an inverse correlation between the amount of precipitation and atmospheric pollution. For example, in summer, rainstorm is the best air purifier from sulfur oxides and nitrogen, and drizzling rain is the best air purifier from other impurities. In winter, there is maximal effect of leaching impurities from the atmosphere by snowflakes because of their large surface area.

In the case of transfer impurities from the source, the impact of rainfall on air purification does not manifest significantly.

L R Sonkin considered a problem of restoration of the background of urban air pollution after its purification by rainfall (Sonkin and Nikolaev 1993; Sonkin et al. 2002). The results revealed that restoration of the background occurred in a certain time, for several hours.

The impact of fog on the content and distribution of impurities in the air was hard and peculiar. Firstly, this was due to the fact that the fog was accompanied by high humidity, in other words, by those meteorological conditions, which contributed to increased air pollution. Secondly, accumulation of impurities from the upstream and downstream layers of air was observed in the fog effect. Finally, the existence of droplets of mist in the atmosphere changed the qualitative composition of the impurities, converting them into more toxic. Therefore, air fog very quickly occurred a reaction of oxidation to SO_3 when contaminating with sulfur dioxide with subsequent formation of sulfuric acid droplets.

Character of variation in concentrations of SO_2, NO_2, and phenol was considered in the fog; in the duration of fog of more than 9 h, the concentration of impurities in phenol increased by 60–70%, the concentration of impurities of SO_2 increased by 80–100%, and the concentration of impurities of NO_2 increased by 40%. All this indicated that fogging was one of the factors leading to the increase of concentration of impurities in the air.

Thus, a lot of studies showed that the level of air pollution was not determined only by the power source and chemical composition of emissions but also by the nature of the dispersion of pollutants in the atmosphere.

6.2.2 *Indicators of the Natural Self-Cleaning Ability of the Atmosphere*

The atmosphere has the self-cleaning capacity. Self-purification of the atmosphere is a partial or complete restoration of the natural composition of the atmosphere due to the removal of impurities caused by natural processes. Rain and snow clean atmosphere due to their absorption capacity, removing dust and water-soluble substances from atmosphere. Plants absorb carbon dioxide and release oxygen, which oxidizes organic matter (the importance of green plants in the self-purification of the atmosphere from carbon dioxide is exceptional. Almost all the free oxygen in the atmosphere has biogenic origin, i.e., green land plants disengage about 30% of free oxygen, and the oceans algae disengage 70% of free oxygen). The ultraviolet rays of the sun kill microorganisms.

Natural self-cleaning capacity of the atmosphere is mainly due to such climatic conditions, as the peculiarities of the underlying surface (vegetation, topography), temperature, amount of precipitation, circulation processes in the atmosphere, and others.

The self-cleaning capacity of the atmosphere also depends on the PAP (potential atmosphere pollution). When the value of the PAP is decreasing, the self-cleaning capacity of the atmosphere increases. It shows how many times as many the average level of air pollution in a particular area ($\overline{q_i}$), caused by the recurrence of adverse weather conditions (inversions, light wind, air stagnation, and fog), is higher than the average level of air pollution in the conventional area ($\overline{q_0}$).

$$\text{PAP} = \frac{\overline{q_i}}{\overline{q_0}} \tag{6.1}$$

As a conditional region, an area where there is minimal repetition of adverse weather conditions is taken. As one city is considered in the work and conditional district is difficult to identify, that hampers the use of this indicator.

One of the main meteorological parameters contributing to the accumulation of harmful substances in the lower atmosphere is a repetition of the wind speed of 0–1 m/s. This process is enhanced by fog; in some cases, it increases the toxicity of impurities. These factors aggravate the atmospheric pollution. It is offered to consider the frequency of days when strong wind and precipitation are the conditions, under which self-purification of air can occur.

It is established that the wind speed, removing the harmful substances from the city, should be at least 6 m/s. The value of precipitations cleaning the atmosphere from pollution is considered of >0.5 mm per day (Perevedentsev and Khabutdinov 2012).

According to Seleguei (1989), meteorological self-cleaning capacity of the atmosphere (K') was the most informative

$$K' = \frac{F_{ws_1} + F_r}{F_{ws_2} + F_{rh}} \tag{6.2}$$

where $F(\%)$ is a frequency, F_{ws_1} is a recurrence of wind speed ≥6 m/s, F_r is a recurrence of precipitation ≥0.5 mm, F_{ws_2} is a recurrence of low wind (0–1 m/s), and F_{rh} is a recurrence of average daily relative humidity ≥80%.

Quantitative evaluation of meteorological conditions for the criterion K' is $K' < 0.8$; these are unfavorable conditions for dispersion; $0.8 \le K' \le 1.4$ are limitedly favorable dispersion conditions; $K' > 1.4$ are favorable conditions for self-cleaning capacity of the atmosphere.

Self-cleaning capacity of the atmosphere takes into account the factors that may impact both air pollution and self-cleaning capacity of the atmosphere. The larger the absolute value of the meteorological potential of self-cleaning capacity of atmosphere K', the better conditions for prevalence of impurities in the atmosphere. If K' is more than 1, then the repeatable processes, contributing to self-cleaning capacity of the atmosphere, dominate over repeatable processes contributing to accumulation of harmful substances therein. If K' is less than 1, then processes contributing to accumulation of contaminants dominate.

According to Perevedentsev and Khabutdinov (2012), during the period of 2013–2016, in Lanzhou City, methodology designed meteorological self-cleaning capacity of the atmosphere (Table 6.3).

The variability of the average annual values of K' was in the range of 1.44–2.27. Annual variations of the coefficient of self-purification K' (Table 6.3) allowed to reveal seasonal patterns of accumulation and prevalence of pollutants. Favorable conditions for the dispersion of pollutants ($K' \ge 1.4$) were observed from February to September. The period of unfavorable conditions for the prevalence of impurities ($K' \le 0.8$) was 3 months (from November to January).

Generally, during the year in Lanzhou, $K' = 2.27$ determined favorable conditions for self-purification of the city's atmosphere.

The repeatability of wind speeds ≥6 m/s was less than the repetition of calms throughout the year, which did not affect the seasonal changes in conditions that determined the processes of self-cleaning of the atmosphere, since the prevailing wind speeds in the city were in the range of 2–4 m/s, which also affected the process

Table 6.3 Monthly average values of the coefficient of self-cleaning capacity of the atmosphere in Lanzhou City (averaging period is 2013–2016)

	2013	2014	2015	2016	Average
January	0.33	0.2	1.25	1.25	0.75
February	1.18	1.25	2.66	2.5	1.89
March	0.42	3	0.66	1.75	1.45
April	2.8	2.6	2.5	11	4.7
May	3.9	6.5	3.5	4.66	4.64
June	2.6	4.7	2.28	6.66	4.06
July	1.1	1.4	2.25	1.57	1.58
August	1.86	1.15	13	1.11	4.28
September	1.15	0.6	1.9	2	1.41
October	1	1.08	2.5	0.64	1.3
November	0.49	0.75	0.33	0.5	0.51
December	0.51	0.5	0.8	0.63	0.61
Average	1.44	1.97	2.8	2.88	2.27

of self-cleaning of the atmosphere (Table 6.4). The conditions of rainfall (intensity of heavy rainfall is higher in the warm season) were higher than in the warm seasons; relative humidity affects the inter-seasonal differences. The combination of all these factors determined the seasonal dynamics of self-purification of atmosphere.

6.2.3 *Correlation Between Meteorological Parameters and Air Pollution*

Contaminants, incoming from different sources into the air basin of the city, affect the microclimatic and mesoclimatic mode of the city and its environs, coupled with orographic features.

To identify the microclimatic and mesoclimatic differences that were manifested in the quantitative values of certain meteorological variables, the data of daily mean daily observations over the period of 2013–2016 years was compared.

Especial interest was in the air temperature distribution in the center of the city and its surroundings, or in the identification of the so-called urban heat island, as well as in the study of wind conditions, which were associated with the prevalence of pollutants within the city.

Influence of Wind Direction on the Distribution of Pollutants

In this paper, the frequency of wind in the four seasons in Lanzhou (Fig. 6.8) was plotted. The northwestern wind and northwestern northerly winds were accounted for 28.7%, while the northern winds and northwestern northerly winds dominated

Table 6.4 Repeatability of (%) of the main meteorological parameters for the calculation of the meteorological self-cleaning capacity of the atmosphere. They are atmospheric precipitation (AP), wind speed (WS), and relative humidity (RH)

	2013				2014				2015				2016			
	WS		AP	RH	WS		AP	RH	WS		AP	RH	WS		AP	RH
	0–1 m/s	≥6 m/s	>0.5 mm	≥80%	0–1 m/s	≥6 m/s	>0.5 mm	≥80%	0–1 m/s	≥6 m/s	>0.5 mm	≥80%	0–1 m/s	≥6 m/s	>0.5 mm	≥80%
January	22	–	9.6	6.4	12.9	–	3.22	3.22	9.6	–	16.1	3.2	6.4	–	16.1	6.4
February	14	3.5	21.4	7.1	10.7	–	35.7	17.8	–	–	28.5	10.7	7.1	–	17.8	–
March	16	–	9.6	6.4	3.22	–	9.6	–	9.6	–	6.4	–	–	3.2	19.3	12.9
April	9.6	9.6	26.6	3.3	10	3.33	50	10	–	–	33.3	13.3	–	3.3	33.3	3.3
May	12.9	3.2	61	3.2	3.22	6.4	35.4	3.22	9.6	3.22	41.9	3.2	3.2	3.2	41.9	6.4
June	13.3	–	53.3	6.6	3.33	–	63.3	10	10	–	53.3	13.3	3.3	–	66.6	6.6
July	32.2	–	67.7	29	12.9	–	51.6	22.5	6.4	–	58	19.3	3.2	–	35.4	19.3
August	16.1	–	41.9	6.4	16.1	–	58	35.4	–	–	41.9	3.2	12.9	–	32.2	16.1
September	23.3	–	50	20	30	–	50	53.3	6.6	–	63.3	26.6	6.6	–	53.3	20
October	16.1	–	29	12.9	12.9	–	41.9	25.8	–	–	32.2	13.9	16.1	–	35.4	38.7
November	20	–	16.6	13.3	20	–	30	20	16.6	–	23.3	53.3	6.6	–	3.3	–
December	25	–	16.1	6.4	6.4	–	3.22	–	3.2	–	12.9	12.9	9.6	–	6.4	–

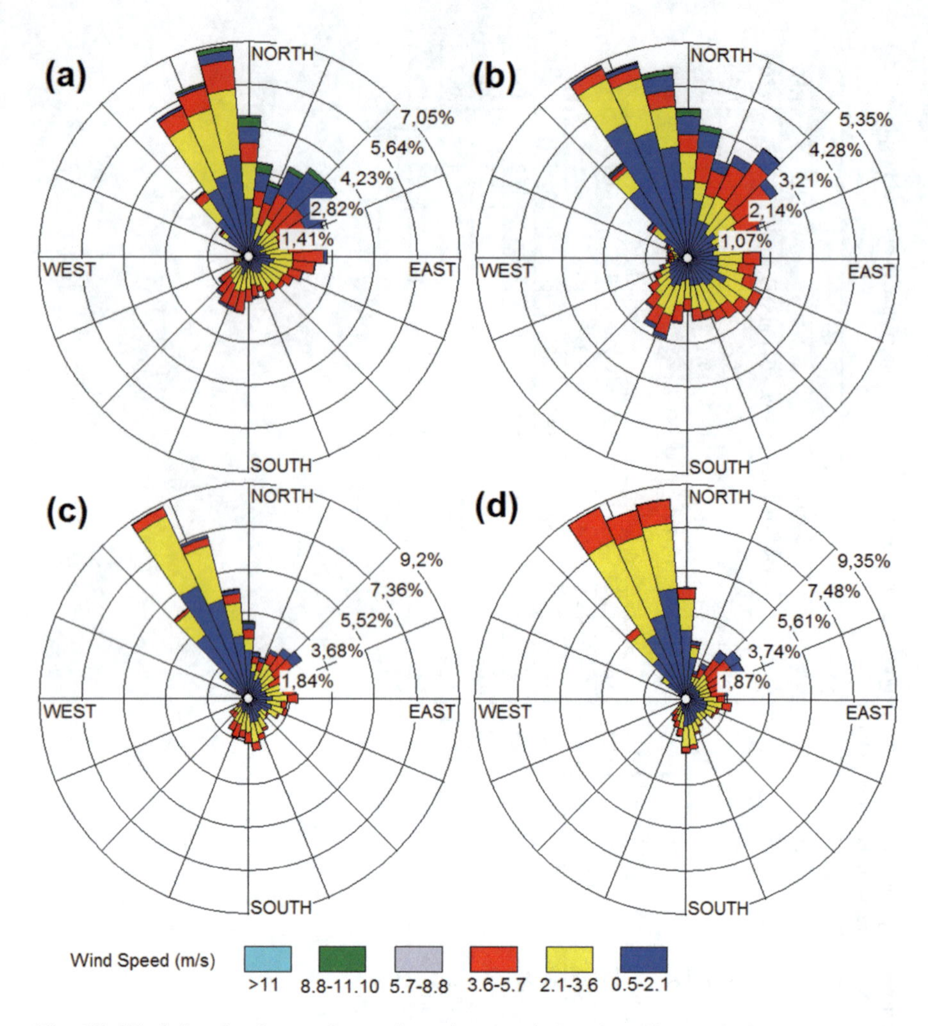

Fig. 6.8 Wind direction frequently rose in each season in Lanzhou City. (**a**) Spring, (**b**) summer, (**c**) autumn, (**d**) winter

in the autumn, and their frequency reached 43.21%; the northwestern wind and the western northwestern wind frequency in winter reached 48.3%. Comprehensive annual situation can be observed in Lanzhou City, when the north and northwestern winds have the dominant wind directions.

Figure 6.9 illustrated that the value of mass concentration of SO_2 was the highest in Lanzhou, the value of an average mass concentration was of 82.4 μg/m^3; followed by northeasterly wind, the value of an average mass concentration was of 75.4 μg/m^3; and the lowest value of concentration was of 38.7 μg/m^3. The highest value was 1.90 times of the lowest value. The wind direction at the highest concentration of NO_2 was also north and northeasterly, with the average mass concentration of 46.9 μg/m^3 and of 41.6 μg/m^3, respectively; and the lowest value of concentration was of 26.4 μg/m^3.

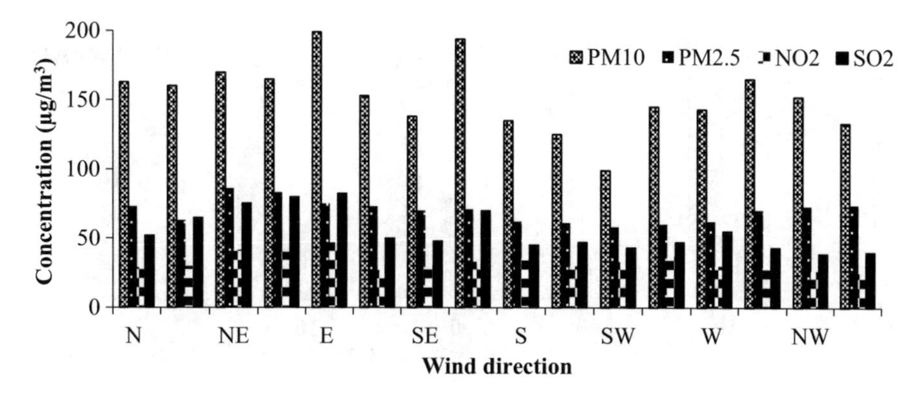

Fig. 6.9 Correlation between concentrations of $PM_{2.5}$, PM_{10}, SO_2, and NO_2 and wind direction in Lanzhou City

In addition, the highest value was 1.56 times of the lowest value. The value of concentration of PM_{10} was the highest at an average value of mass concentration of 199 μg/m^3; followed by south-southeast wind, the average value of mass concentration was of 194 μg/m^3; when the southeast wind blew, the mass concentration was the lowest, its average value was of 87 μg/m^3, and the highest value was 1.90 times of the lowest value. Maximum concentration of $PM_{2.5}$ was observed under the northeast wind, it was of 86 μg/m^3, and under the southwest wind, value of concentration was of 58 μg/m^3; it was the minimum one.

In general, the concentration of pollutants in eastern or southeastern winds in Lanzhou was higher, which was not conducive to the diffusion of air pollutants. The concentration of pollutants in northwestern was lower, which was favorable for the diffusion of pollutants. There might be three reasons. They were the following:

1. In Lanzhou City, the frequency of the summer northwestern wind was relatively high, while the degree of air pollution in summer was relatively low compared with other seasons.
2. When the northwestern wind was blowing, the wind speed was high; often, cold air transport, diffusion ability enhancement, and dilution and diffusion of pollutants were more significant.
3. When the wind blew in Lanzhou, the region might not be consistent with a large range of winds and urban areas.

Impact of Temperature on the Concentration of Pollutants in Lanzhou City

The average mass concentration of SO_2 was of 41 μg/m^3, the average mass concentration of NO_2 was of 46.6 μg/m^3, and the average mass concentration of PM_{10} was of 125 μg/m^3, respectively, when the temperature was below −10 °C; when the temperature was −10~0 °C, the average mass concentration of SO_2 was of 35 μg/m^3, the

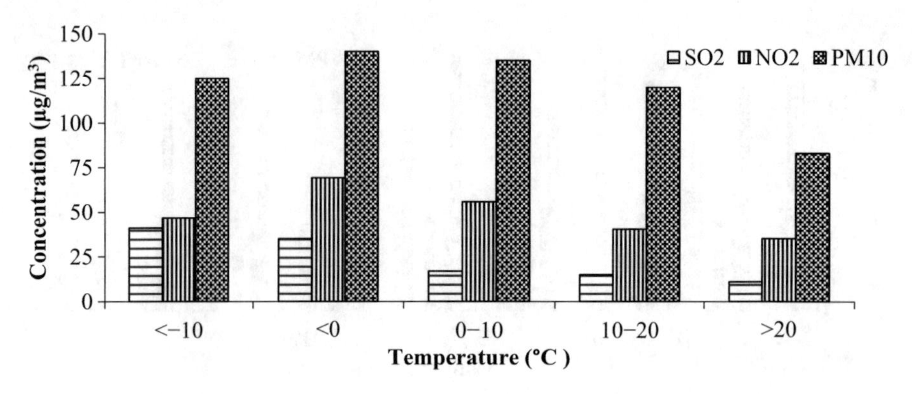

Fig. 6.10 Correlation between concentrations of SO_2, NO_2, and PM_{10} and temperature in Lanzhou City

average mass concentration of NO_2 was of 69.3 μg/m³, and the average mass concentration of PM_{10} was of 140 μg/m³, respectively; when the temperature was 0~10 °C, respectively, the average mass concentration of SO_2 was of 17 μg/m³, the average mass concentration of NO_2 was of 55.8 μg/m³, and the average mass concentration of PM_{10} was of 135 μg/m³; when the temperature was 10~20 °C, the average mass concentration of SO_2 was of 15 μg/m³, the average mass concentration of NO_2 was of 40.3 μg/m³, and the average mass concentration of PM_{10} was of 120 μg/m³, respectively; when the temperature exceeded 20 °C, respectively, the average mass concentration of SO_2 was of 11 μg/m³, the average mass concentration of NO_2 was of 32.5 μg/m³, and the average mass concentration of PM_{10} was of 83 μg/m³, respectively (Fig. 6.10). On the one hand, the trend of this distribution of mass concentration was because when the average temperature was below 10 °C, the heating period began in Lanzhou region, and coal and other fuel consumption increased significantly; on the other hand, the trend of this distribution of mass concentration might be because the ground temperature was relatively low, and convection was weak and easy to form inversion.

Figure 6.11 illustrates that the obvious dependence of concentrations of $PM_{2.5}$ and O_3 on temperature is visible between throughout a year. Since the temperature changed of less than −10 °C to >20 °C, the average concentrations of O_3 changed at temperatures above 10 °C to 75 μg/m³, and the average concentrations of O_3 changed at temperatures below 0 °C to 67 μg/m³, at temperatures of 0–10 °C to 48 μg/m³, and at temperatures of 10–20 °C to 38 μg/m³, and they exceed temperature of 20 °C to 35 μg/m³. Opposing indicators of concentrations of O_3 are showed, where the lowest concentrations were at a minimum temperature and where they gradually increased to a maximum value at high positive temperatures, so at temperature < −10 °C, concentration of O_3 was of 25 μg/m³, and at temperature > 20 °C, concentration of O_3 was of 58 μg/m³.

Ground temperature has no significant impact on the concentration of pollutants; temperature is the heating needs. The lower the temperature, the higher the demand for heating, the more pollutants will be emitted. However, in the temperature range

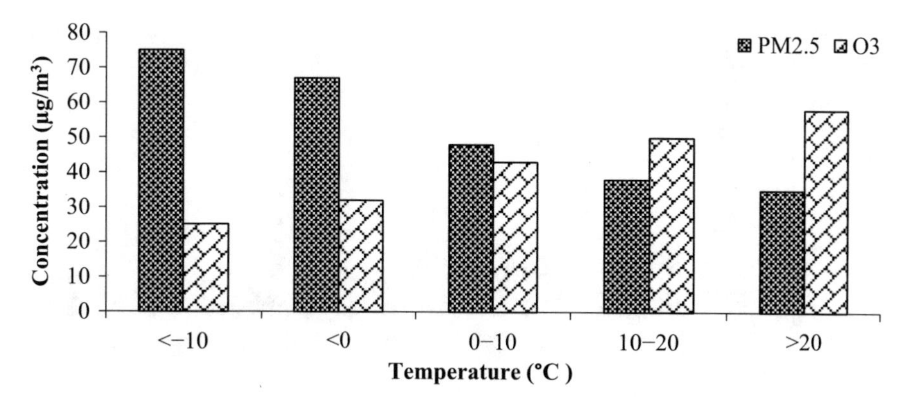

Fig. 6.11 Correlation between concentrations of $PM_{2.5}$ and O_3 and temperature in Lanzhou City

below −10 °C, −10~0 °C, and 0~10 °C, concentrations of the three pollutants (SO_2, NO_2, and PM_{10}) were the highest at the temperature of −10~0 °C; the lower the temperature, the higher the average mass concentration of pollutants; the analysis revealed that there were two reasons for this situation. They were the following:

1. The temperature was in the range of −10~0 °C, and the average precipitation was 0.0778 mm; when the temperature was less than −10 °C, the average precipitation was 0.5236 mm; it was 6.73 times as much the former. This form of precipitation should be snow, and snow purification of air pollutants was very strong, so the temperature of the concentration of pollutants was less than −10 °C range than the −10~0 °C.
2. The average wind speed was 0.66 m/s when the temperature was in the range of −10~0 °C, and the average wind speed was 1.757 m/s when the temperature was less than −10 °C. This indicated that if the temperature was less than −10°C when the wind speed was significantly higher than with temperature −10~0°C, the dilution and purification ability of air pollutants were strong, so the mass concentration of pollutants with temperature > −10°C is lower than −10~0°C. Inversion temperature was between −10~0 °C in the most stable layer, and when inversion temperature was below −10 °C, there were usually a cold air activity, stratification instability, and the ability to diffuse in the atmosphere; if it was accompanied by snow, air pollution had a strong purifying effect.

The Influence of the Amount of Precipitations on the Concentration of Major Pollutants in the City

Air pollution in Lanzhou was mainly caused by the combustion of fuel. Most pollutants presented in the atmosphere as aerosol particles. Precipitation had obvious purifying effect on air pollutants. Figure 6.12 illustrated that the average mass concentration of SO_2 was of 85.8 μg/m³ on the nonprecipitation day; and the average mass concentration of SO_2 was of 47.9 μg/m³ and less than 5 mm on the precipitation

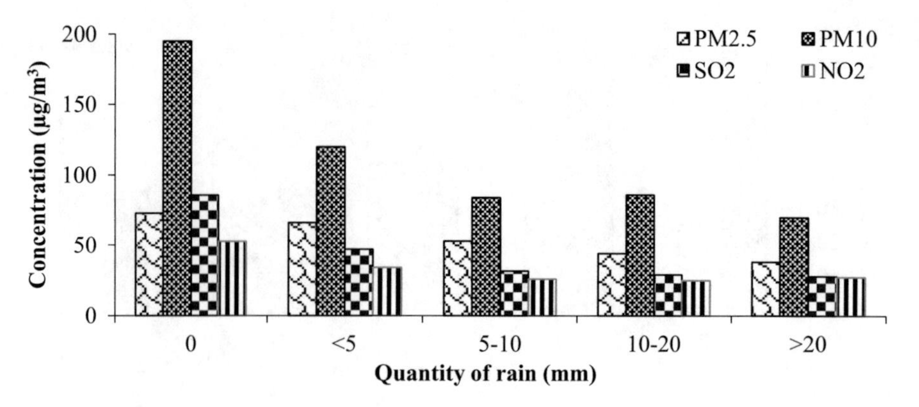

Fig. 6.12 Correlation between concentrations of PM_{10}, $PM_{2.5}$, SO_2, and NO_2 and precipitation in Lanzhou City

day; the average mass concentration of SO_2 decreased by 44.2%; and the average mass concentration of SO_2 was of 31.6 μg/m³ at 5~10 mm, which was 63.2% less than that on nonprecipitation days. If precipitation was of 10~20 mm and if precipitation was more than 20 mm, the average mass concentrations of SO_2 were of 29.16 μg/m³ and of 28.0 μg/m³, decreasing by 66.08% and by 67.36%, respectively.

The average daily concentration of NO_2 was of 52.8 μg/m³ on nonprecipitation days; and it was of 34.2 μg/m³ on days when precipitation did not exceed 5 mm, and concentration decreased by 35.22%; and it decreased by 51% when the concentration was of 25.9 μg/m³ if precipitation was from 5 to 10 mm per day. If an average daily precipitation was of 10~20 mm and more than 20 mm, the mean daily concentrations were of 25 μg/m³ and of 27.3 μg/m³, respectively, which was 52.65% and 48.29% less than on nonprecipitation days.

On nonprecipitation days, the average concentrations of $PM_{2.5}$ were of 73.3 μg/m³; if precipitation was of less than 5 mm, the average concentrations of $PM_{2.5}$ were of 66.2 μg/m³, decreasing by 9.66%. If precipitation was of 5–10 mm, the average concentrations of $PM_{2.5}$ were of 53.1 μg/m³, which was 27.55% less than that on nonprecipitation days; when precipitation was of 10–20 mm and 20 mm or more, the average concentration of $PM_{2.5}$ was of 44 μg/m³ and of 38.5 μg/m³, respectively, which was 40% and 47.47% less than that on nonprecipitation days.

The average mass concentration of PM_{10} was of 195 μg/m³ on a nonprecipitation day; and average mass concentration of PM_{10} was of 120.2 μg/m³ if precipitation was less than 5 mm; it decreased by 37.39%. The average mass concentration of PM_{10} was of 84.8 μg/m³ if precipitation was of 5~10 mm, which was 56.51% less than that on nonprecipitation days. The mean mass concentrations of PM_{10} were of 86.3 μg/m³ and of 70.4 μg/m³ if precipitation was of 10~20 mm and if precipitation was more than 20 mm, respectively, which decreased by 55.74% and 63.89%, respectively.

Atmospheric environment pollutants had a background mass concentration value; the impact of precipitation on pollutants was obvious when it was much greater than this background mass concentration value, but the impact of precipitation on the pollutants was not obvious when the value was close to this background mass concentration value. Under identical conditions of precipitation, concentrations of SO_2 were more sensitive to purification than other pollutants. Thus, if an average daily rainfall was of 0–5 mm, concentrations of SO_2 decreased by 44.2%; if concentrations of NO_2 decreased by 35.22%, the reason for this situation was that SO_2 was more readily soluble in water than NO_2 during precipitation to land.

The Impact of Wind Speed on the Concentration of Contaminants

Wind is of importance in the dilution and diffusion of air pollutants. The average mass concentration of SO_2 was of 101.4 μg/m^3, the average mass concentration of NO_2 was of 76.5 μg/m^3, and the average mass concentration of PM_{10} was of 209.5 μg/m^3, respectively, which was the most unfavorable condition for pollutants' diffusion. The average mass concentrations of the 3 pollutants were of 43.1 μg/m^3, of 47.3 μg/m^3, and of 139.1 μg/m^3, respectively, decreasing by 57.4%, by 38.17%, and by 33.6%, respectively, compared with the static wind days when the wind speed was less than 1.5 m/s; when the wind speed was more than 1.5 m/s, the average concentrations of the 3 pollutants were of 21.3 μg/m^3, of 35.8 μg/m^3, and of 108.7 μg/m^3, decreasing by 78.99%, by 53.2%, and by 48.11%, respectively. That indicated that the higher the wind speed, the stronger the ability of the dilution and diffusion of pollutants. In calm weather, average concentration of $PM_{2.5}$ was of 83.1 μg/m^3; at wind speeds below 1.5 m/s, average concentration of $PM_{2.5}$ was of 61.5 μg/m^3, decreasing by 25.99% regarding calm weather; at wind speeds above 1.5 m/s, the concentration of $PM_{2.5}$ was of 33.6 μg/m^3, decreasing by 59.6% regarding calm weather. These are the following reasons of the most obvious dilution effect of increasing wind speed on SO_2, $PM_{2.5}$, and NO_2:

1. $PM_{2.5}$, SO_2, and NO_2 are gaseous pollutants, and PM_{10} is a particulate pollutant; it is obvious that under the same wind speed conditions, the gaseous substances dilute the diffusion easier than the particulate matter.
2. $PM_{2.5}$, SO_2, and NO_2 mainly originate from the source of industrial enterprises and the source of life, while PM_{10} mainly originate from industrial enterprise source and source of life, as well as from dust sources. On the other hand, PM_{10} emission was enhanced by external transportation and local secondary dust, but the concentration of PM_{10} was decreased as a whole. The PM_{10} emission was enhanced by PM_{10} emission from local industrial enterprises.

During the period of statistics, the static wind days accounted for 11.39%, the small wind days less than 1.5 m/s accounted for 68.78%, while the days with more than 1.5 m/s accounted for only 19.82%, less than one fifth of the total (Fig. 6.13). That was because, on the one hand, there was Lanzhou in the valley basin, surrounded by mountains, affecting the local atmospheric circulation; on the other hand, there were significantly increased height and density of the buildings under the acceleration of urbanization, increased surface roughness, and reduced surface wind speed.

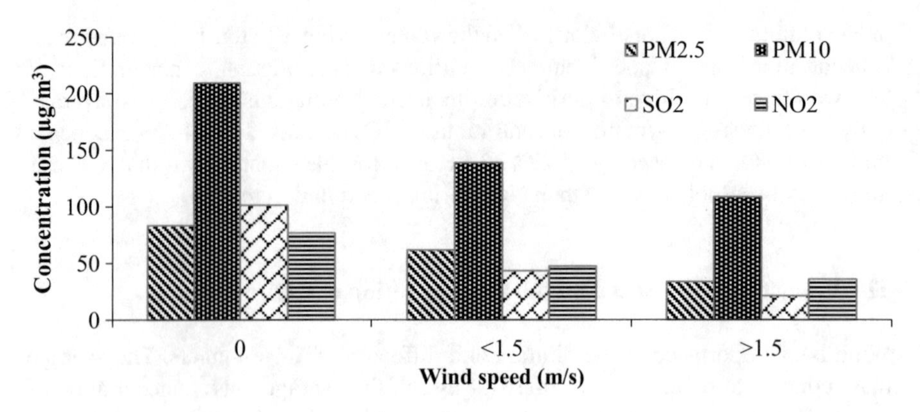

Fig. 6.13 Correlation between concentrations of PM_{10}, $PM_{2.5}$, SO_2, and NO_2 and wind speed in Lanzhou

Table 6.5 Pearson's correlation coefficients (*r*) between daily meteorological factors and air pollutants for Lanzhou (period 2013–2016)

	$PM_{2.5}$	PM_{10}	SO_2	NO_2	CO_2	O_3
T_{ave}	−0.411	−0.209	−0.657	−0.287	−0.617	0.488
T_{min}	−0.399	−0.201	−0.651	−0.271	−0.609	0.458
T_{max}	−0.396	−0.2	−0.622	−0.286	−0.587	0.491
P_{ave}	0.184	0.049	0.234	0.179	0.274	−0.309
RH	−0.042	−0.083	−0.117	0.037	−0.017	−0.06
WS	−0.035	0.066	−0.047	−0.005	−0.111	0.163
V	−0.223	−0.074	−0.37	−0.208	−0.343	0.277

6.2.4 Correlation Between Climatic Factors and Air Pollutants

Correlation between impurity concentrations was conducted to determine the correlation between climatic factors and air pollution in cities in the pool with a variety of meteorological parameters (average temperature (T_{ave}, °C), minimum temperature (T_{min}, °C), maximum temperature (T_{max}, °C), pressure (P_{ave}, hPa), wind speed (WS, m/s), relative humidity (RH, %), and visibility (*V*, km)) (Tables 6.5 and 6.6).

Among all above parameters (temperature, wind, humidity, visibility, and atmospheric pressure), more close correlation between the concentration of impurities and air temperature was observed. During warmer months, there was a positive correlation between the temperature and the concentrations of $PM_{2.5}$, SO_2, and CO.

Increasing summer temperatures were often observed in anticyclone type of weather conditions, which was accompanied by a weak or quiet wind and the lack of precipitation, i.e., by the conditions under which the impurities accumulate in the atmosphere. Moreover, in these cases, the rate of photochemical reactions increased, at which content of secondary air pollutants also increased. Dust, raised by motor transport, as well as particulate matter (PM_{10}) emitted by industrial sources, was

Table 6.6 Pearson's correlation coefficients (r) between seasonal meteorological factors and air pollutants for Lanzhou (2013–2016)

	Winter						Spring					
	$PM_{2.5}$	PM_{10}	SO_2	NO_2	CO	O_3	$PM_{2.5}$	PM_{10}	SO_2	NO_2	CO	O_3
T_{ave}	−0.182	0.094	−0.461	−0.259	−0.326	0.434	−0.013	−0.071	−0.059	0.069	−0.18	0.189
T_{min}	−0.153	0.141	−0.462	−0.222	−0.347	0.446	0.040	−0.012	−0.069	0.068	−0.184	0.189
T_{max}	−0.147	0.03	−0.332	−0.205	−0.21	0.311	−0.05	−0.106	−0.037	0.06	−0.145	0.151
P_{ave}	0.18	−0.005	0.235	0.264	0.196	−0.332	0.04	0.031	−0.107	−0.079	−0.065	0.01
RH	0.029	−0.006	0.048	0.0851	0.028	−0.028	−0.042	−0.026	−0.15	−0.048	−0.187	0.258
WS	−0.117	0.02	−0.218	−0.035	−0.178	0.152	0.102	0.088	−0.03	0.017	0.011	0.061
V	−0.099	−0.014	−0.186	−0.064	−0.174	0.15	0.067	0.048	−0.097	−0.072	−0.018	0.079
	Summer						Autumn					
T_{ave}	0.086	−0.007	0.064	−0.022	0.074	0.012	−0.442	−0.249	−0.61	−0.378	−0.54	0.488
T_{min}	0.031	−0.012	0.149	−0.004	0.08	0.017	−0.454	−0.247	−0.603	−0.345	−0.557	0.507
T_{max}	0.09	0.004	−0.02	−0.021	0.045	0.016	−0.382	−0.221	−0.547	−0.357	−0.46	0.402
P_{ave}	−0.17	0.013	−0.124	−0.011	−0.028	0.11	0.029	0.006	0.085	−0.010	0.026	−0.049
RH	−0.036	−0.049	0.018	0.043	0.083	−0.11	−0.075	0.066	−0.062	0.094	−0.123	0.058
WS	−0.079	−0.119	−0.022	−0.112	−0.11	0.142	−0.09	−0.078	−0.094	0.012	−0.087	0.089
V	−0.002	0.028	0.013	0.016	−0.026	0.037	−0.189	−0.203	−0.293	−0.132	−0.192	0.238

stored in the airspace of the cities in dry hot weather for a long time. In the cold season, weak negative correlation and moderate negative correlation between air temperature and concentrations of nitrogen oxides, sulfur, and carbon were observed. In this case, concentration increased, and the anthropogenic factor (fuel combustion increases) and adverse weather conditions (no wind in an anticyclone, temperature inversion, freezing fog, etc.) increased.

The wind speed mainly impact on the concentrations of $PM_{2.5}$, PM_{10}, NO_2, and CO, formaldehyde and dust, a major source of which is transport. Reverse nature of the correlation indicated that wind speeds decreased air pollution from low sources. In general, the contribution of the change concentration of impurities in wind speed was 35–40%.

The character of correlation between air pollution indices and humidity was very unstable. The correlation coefficients between the concentrations of impurities and humidity were negative and close in magnitude with correlation coefficients between the pollutants and air temperature. That confirmed only the dependence of these characteristics on moisture because of air temperature. The most significant negative correlation was observed in spring with the concentrations of oxides of sulfur, nitrogen, and carbon. During this period, at a high content of water vapor in the atmosphere, chemical reactions occurred between SO_2 and NO_2 and water vapor (H_2O) and resulted in formation of more complex compounds (H_2SO_4, HNO_3, etc.), and the impurity concentration might reduce by themselves.

Under atmospheric pressure, the correlation was positive mainly in autumn and in winter, i.e., when the pressure was increasing, the air pollution increased.

Thus, there is the correlation between the concentration of pollutants in the air of the Lanzhou City and different meteorological parameters, but level of correlation varies much depending on season. Between closest correlation observed in spring and in autumn, the impact of the anthropogenic factor was reduced, and the weather conditions were the most unstable.

Chapter 7
Conclusions and Prospects

7.1 Conclusions

Studies of air pollution, as well as of the way of pollutants entering the city in the period 2003–2016, were received basing on the monitoring of air quality data obtained from the Ministry of Environmental Protection of the People's Republic of China, basing on satellite's data of MODIS and Landsat 8, as well as on model-simulated data, which allowed us to draw the following conclusions:

1. In a region, average concentrations of PM exceeded the World Health Organization guideline values, which indicated a serious impact of concentrations on people and the environment of Lanzhou City. Seasonal and diurnal course of pollutants and pollution probabilities were characterized, respectively. They were characterized as the maximum in the winter and summer months in the afternoon. The annual course was largely determined by the seasonal mode of operation of the heat and power unit and vehicles. In summer months, the dependence of the increase in the concentrations of secondary pollutants on the rate of photochemical processes in the atmosphere was revealed. Daily variations in concentrations were associated with atmospheric turbulence.
2. Seasonal characteristic of air quality in Lanzhou had obvious differences. According to the seasonal distribution, the average concentration of $PM_{2.5}$, of SO_2, of NO_2, and of CO quarter value had the same dynamics, i.e., winter > autumn > spring > summer. The mass concentration value of PM_{10} varied from heavy to light in seasons, winter > spring > autumn > summer, while difference between winter and spring concentration values was not high; spring dust weather increased concentration of PM_{10}; however, strong wind was of importance in the removal of gaseous pollutants and $PM_{2.5}$; thus the concentrations of $PM_{2.5}$, of SO_2, of NO_2, and of CO in spring were a little lower than those in autumn.
3. The concentrations of all pollutants (in addition to the properties of their sources) significantly depended on the complex of meteorological quantities. It was

M. Filonchyk, H. Yan, *Urban Air Pollution Monitoring by Ground-Based Stations and Satellite Data*, https://doi.org/10.1007/978-3-319-78045-0_7

necessary to attribute the following conditions to unfavorable meteorological conditions of dispersion of impurities in Lanzhou; on the surface of the Earth, unfavorable meteorological conditions of dispersion of impurities were the following: quiet or low wind, prevailing temperature gradients and wind direction, relative air humidity exceeding 60%, fogs and hazes at the height of the leading stream of 3 km, and speeds of west and northwest winds of 9–19 m/s. The range of increased concentrations of nitrogen oxides, maximum of sulfur, and carbon containing in urban atmosphere was mainly related to meteorological parameters. This correlation allowed us to construct a multidimensional regression model of atmospheric pollution.

4. Atmospheric pollution did not depend only on a complex of meteorological quantities, but it also affected the thermodynamic state of the air, creating "heat islands," as a feedback. The data obtained from MODIS and Landsat 8 satellites revealed that microclimatic measurements allowed to register increasing air temperature in the city center and in the industrial zone by 1–3 °C comparing to its margins. "Heat island" led to an upward movement of air in the city center during the day, which helped to reduce the concentration of impurities near the surface of the Earth, to some extent. The nature of this phenomenon is related to both the direct release of heat into the atmosphere by municipal services and the radiation heating of the polluted air. The NDVI index indicated a lack of vegetation in urban area.
5. Aerosol optical depth data indicated that the environmental situation of the city had tended to improve, in comparison with other regions of the country, but it had still been at a level indicating the need for continuous monitoring of aerosols in the atmosphere of the city. Maximum aerosol contamination was observed in spring, due to sandstorms at the current time, indicating a high level of concentration of natural aerosols. The minimum one was registered in the summer-autumn period.
6. Except for local pollution sources, there are external sources of pollution located outside the study area and associated with passage of contaminants from elevated and desert regions of Western and Northern China, which are based on CALIPSO, OMI, NAAPS, and HYSPLIT and have three main trajectories. They are "north," associated with air masses passing through the Gobi Desert, Republic of Mongolia, and Inner Mongolia; "northwest," appearing mainly in the northwestern part of Xinjiang (Gurbantunggut Desert), where air mass go through Inner Mongolia and Ningxia to Lanzhou City; and "western," originating in the western deserts (Taklamakan Desert in Xinjiang) and in the desert of Qaidam Basin, where air masses pass through the Tibetan Plateau and Qinghai.
7. Studies of the dynamics of air pollution index (API in 2001–2012 and AQI in 2013–2016) in Lanzhou City show a tendency to reduce the level of pollution. The API value decreased continuously since 2001–2005; it increased suddenly in 2006. In 2007, the API value was the lowest in 11 years, accounting for 90, air quality achieved the best value, and then the API value increased year by year; by 2011 it fell to 95. From the API average, 2007, 2008, and 2011 were the best years of air quality in the last 11 years; the API average was 90–100. In 2006,

there was the most serious pollution than in 6 years before and 6 years after, and the API standard deviation also reached the highest value. In the period of 2013–2016, the AQI gradually decreased and reached the lowest value in 16 years.

7.2 Contributions

Though some of the previous research studied air pollution in the city, they were not enough to determine the main pollution sources and causes, because they were researched for a short period and a special kind of data was used. This book was focused on studying multiseasonal characteristics of air pollution and spatial and temporal characteristics using comprehensive approach. The scientific novelty of this work consists in the following:

1. Assessing the current level of air pollution, it is necessary to know the value of the concentration of pollutants in the air in order to reduce or to increase this concentration in the atmosphere. Following this, a method for forecasting the level of atmospheric pollution, taking into account physical and geographical conditions, classes of synoptic situations, seasons, and local features allowing to improve the quality of assessment of the ecological state of the air basin and developing effective measures to protect atmospheric purity, has been developed.
2. Taking into account changes in the microclimate, the principles of a comprehensive assessment of the impact of a city on the air basin were developed both locally and regionally.
3. It was justified range of impurity concentrations as an objective criterion for the level of air pollution in the cities and to identify trends of their temporal and spatial variations.
4. Firstly, it was zoning of the territory in terms of the natural self-cleaning capacity of the atmosphere; weather conditions, contributing to increased air pollution in Lanzhou City, were determined.
5. Principles of integrated assessment of air pollution, formation sources of air pollutants, and the temperature of the underlying surface were developed by using the GIS technologies.

7.3 Recommendations for Further Research

In this work, Lanzhou City and the main air pollutants are the targets of research, and as the research mechanism for regional monitoring of air pollutants is very complex, it is very important to study the dynamic monitoring of regional atmospheric environmental quality by extracting the atmospheric radiation components from the satellite remote sensing image data, which is of great importance to monitor and to forecast regional atmospheric environmental quality. The collection and analysis of data and the use of the technical means for research affect many related disciplines. The author believes that several of the following aspects deserve further in-depth study.

Further study of the environmental situation should not be carried out only in one concrete place but also in a complex of the territories of a whole region or a country. Later, the research will allow to compare the data for different regions and to find out the sources of pollution located outside the study area.

A comprehensive study of atmospheric pollution is based on satellite data, and ground monitoring data is required. Further integrated use of the obtained data can provide more accurate and reliable information on the levels of pollution, on the nature of the origin, and on the ways of transport of pollutants.

The research of pollution sources of Lanzhou City, which included industrial, motor vehicles, natural aerosol, and other emissions of atmospheric and chemical properties, was carried out for combination source analysis technology to identify the contribution of the sources of pollution, the secondary fine particles, and photochemical pollution and the characteristics of the precursor differences and mutual influence, to facilitate better development of air pollution control measures, to further improve the quality of atmospheric environment in Lanzhou City, to protect the health of local citizens, and to provide scientific and technological support.

The time of pollution changes and pollution levels are limited. Air pollutants and the level of pollution are changing under the development of economic and social factors. The long-term and continuous monitoring results are continuously studied and improved.

Further verification of temporal and spatial distribution of pollutants. The spatial and temporal distribution of atmospheric particulate matter in Lanzhou was simulated by spatial interpolation method. The accuracy of the interpolation results was verified by the method of recross validation, which could be coupled with a certain period of inversion results obtained from satellite, for further verifying the accuracy of its forecast results.

References

Abdou, W. A., Diner, D. J., Martonchik, J. V., Bruegge, C. J., Kahn, R. A., & Gaitley, B. J. (2005). Comparison of coincident Multiangle Imaging Spectroradiometer and Moderate Resolution Imaging Spectroradiometer aerosol optical depths over land and ocean scenes containing Aerosol Robotic Network sites. *Journal of Geophysical Research Atmospheres, 110*(D10), 1275–1287.

Abelsohn, A., & Stieb, D. M. (2011). Health effects of outdoor air pollution approach to counseling patients using the Air Quality Health Index. *Canadian Family Physician, 57*(8), 881–887.

Abuduwailil, J., Zhaoyong, Z., & Fengqing, J. (2015). Evaluation of the pollution and human health risks posed by heavy metals in the atmospheric dust in Ebinur Basin in Northwest China. *Environmental Science and Pollution Research, 22*(18), 14018–14031.

Aili, A., & Oanh, N. T. K. (2015). Effects of dust storm on public health in desert fringe area: Case study of northeast edge of Taklimakan Desert, China. *Atmospheric Pollution Research, 6*(5), 805–814.

Atkinson, R. W., Ross, A. H., Sunyer, J., Ayres, J., & Baccini, M. (2001). Acute effects of particulate air pollution on respiratory admissions: Results from APHEA 2 project. *American Journal of Respiratory and Critical Care Medicine, 164*(10), 1860–1866.

Ayala, A., Brauer, M., Mauderly, J. L., & Samet, J. M. (2012). Air pollutants and sources associated with health effects. *Air Quality, Atmosphere and Health, 5*(2), 151–167.

Beckett, K. P., Freersmith, P. H., & Taylor, G. (2000). Particulate pollution capture by urban trees: Effect of species and wind speed. *Global Change Biology, 6*(8), 995–1003.

Beelen, R., Hoek, G., Brandt, P. A. V. D., Goldbohm, R. A., Fischer, P., & Schouten, L. J. (2008). Long-term effects of traffic-related air pollution on mortality in a Dutch cohort (NLCS-AIR study). *Environmental Health Perspectives, 116*(2), 196–202.

Bench, G., Fallon, S., Schichtel, B., Malm, W., & Mcdade, C. (2007). Relative contributions of fossil and contemporary carbon sources to PM 2.5 aerosols at nine. Interagency Monitoring for Protection of Visual Environments (IMPROVE) network sites. *Journal of Geophysical Research Atmospheres, 112*(D10), 326–330.

Cao, J. J., Wu, F., Chow, J. C., & Lee, S. C. (2005). Characterization and source apportionment of atmospheric organic and elemental carbon during fall and winter of 2003 in Xi'an, China. *Atmospheric Chemistry and Physics, 5*(11), 3127–3137.

Carrer, D., Meurey, C., Ceamanos, X., Roujean, J. L., Calvet, J. C., & Liu, S. (2014). Dynamic mapping of snow-free vegetation and bare soil albedos at global 1km scale from 10-year analysis of MODIS satellite products. *Remote Sensing of Environment, 140*(1), 420–432.

Chen, Y. Z., & Xiao, H. Y. (2009). Classification of atmospheric aerosol and its hazards to human health. *Jiangxi Science, 27*(6), 912–915. (in Chinese).

M. Filonchyk, H. Yan, *Urban Air Pollution Monitoring by Ground-Based Stations and Satellite Data*, https://doi.org/10.1007/978-3-319-78045-0

Chen, Y. W., & Yang, Y. Q. (2001). The research on eclampsia and meteorological conditions. *Gansu Meteorology, 19*(3), 41–44. (in Chinese).

Choi, Y., Park, R. J., & Ho, C. (2009). Estimates of ground-level aerosol mass concentrations using a chemical transport model with Moderate Resolution Imaging Spectroradiometer (MODIS) aerosol observations over East Asia. *Journal of Geophysical Research Atmospheres, 114*(D4), 83–84.

Christensen, J. H. (1997). The Danish eulerian hemispheric model – a three-dimensional air pollution model used for the arctic. *Atmospheric Environment, 31*(24), 4169–4191.

Christopher, S. A., & Gupta, P. (2010). Satellite remote sensing of particulate matter air quality: The cloud-cover problem. *Journal of the Air and Waste Management Association, 60*(5), 596–602.

Chu, D. A., Kaufman, Y. J., Zibordi, G., Chern, J. D., Mao, J., & Li, C. (2003). Global monitoring of air pollution over land from the Earth Observing System-Terra Moderate Resolution Imaging Spectroradiometer (MODIS). *Journal of Geophysical Research: Atmospheres, 108*(D21), 4661.

Chu, P. C., Chen, Y., Lu, S., Li, Z., & Lu, Y. (2008). Particulate air pollution in Lanzhou China. *Environment International, 34*(5), 698–713.

Corn, M., Stein, F., Hammad, Y., Manekshaw, S., Freedman, R., & Hartstein, A. M. (1973). Physical and chemical properties of respirable coal dust from two United States mines. *American Industrial Hygiene Association Journal, 34*(7), 279–285.

Darlington, T. L., Kahlbaum, D. F., Heuss, J. M., & Wolff, G. T. (1997). Analysis of PM10 trends in the United States from 1988 through 1995. *Journal of the Air and Waste Management Association, 47*(10), 1070–1078.

Die, H. U., Zhang, L., Sha, S., & Wang, H. (2013). Contrast and Application of MODIS Aerosol Products over the Arid and Semiarid Region in Northwest China. *Journal of Arid Meteorology, 31*(4), 677–683. (in Chinese).

Dimitriou, K., Paschalidou, A. K., & Kassomenos, P. A. (2017). Distinct synoptic patterns and air masses responsible for long-range desert dust transport and sea spray in Palermo, Italy. *Theoretical and Applied Climatology, 130*(3-4), 1123–1132.

Diner, D. J., Braswell, B. H., Davies, R., Gobron, N., Hu, J., & Jin, Y. (2005). The value of multiangle measurements for retrieving structurally and radiatively consistent properties of clouds, aerosols, and surfaces. *Remote Sensing of Environment, 97*(4), 495–518.

Ding, F., & Hanqiu, X. U. (2006). Comparison of two new algorithms for retrieving land surface temperature from landsat TM thermal band. *Geo-Information Science, 8*(3), 125–130.

Dockery, D. W., & Pope, C. A. (1994). Acute respiratory effects of particulate air pollution. *Annual Review of Public Health, 15*(1), 107–132.

Downs, S. H., Schindler, C., & Liu, L. J. S. (2007). Reduced exposure to PM10 and attenuated age-related decline in lung function. *New England Journal of Medicine, 357*(23), 2338–2347.

Dubovik, O., Herman, M., Holdak, A., Lapyonok, T., Tanré, D., & Deuzé, J. L. (2011). Statistically optimized inversion algorithm for enhanced retrieval of aerosol properties from spectral multi-angle polarimetric satellite observations. *Atmospheric Measurement Techniques, 4*(5), 975–1018.

Durkee, P. A., Jensen, D. R., Hindman, E. E., & Haar, T. H. V. (1986). The relationship between marine aerosol particles and satellite-detected radiance. *Journal of Geophysical Research: Atmospheres, 91*(D3), 4063–4072.

Elminir, H. K. (2005). Dependence of urban air pollutants on meteorology. *Science of the Total Environment, 350*(1–3), 225–237.

Engel-Cox, J. A., Hoff, R. M., Rogers, R., Dimmick, F., Rush, A. C., & Szykman, J. J. (2006). Integrating lidar and satellite optical depth with ambient monitoring for 3-dimensional particulate characterization. *Atmospheric Environment, 40*(40), 8056–8067.

EPA (U.S. Environmental Protection Agency). (2016). *Technical assistance document for the reporting of daily air quality – The Air Quality Index (AQI) (EPA-454/B-16-002)*. Office of Air Quality Planning and Standards. Retrieved from https://www3.epa.gov/airnow/aqi-technical-assistance-document-may2016.pdf

Erp, A. M. V., Kelly, F. J., Demerjian, K. L., Iii, C. A. P., & Cohen, A. J. (2012). Progress in research to assess the effectiveness of air quality interventions towards improving public health. *Air Quality, Atmosphere and Health, 5*(2), 217–230.

Fairley, D. (1990). The relationship of daily mortality to suspended particulates in Santa Clara County, 1980-1986. *Environmental Health Perspectives, 89*(4), 159–168.

Fan, S. M. (2013). Modeling of observed mineral dust aerosols in the arctic and the impact on winter season low-level clouds. *Journal of Geophysical Research Atmospheres, 118*(19), 161–174.

Fang, X., Yun, X., & Li, L. (2003). Effects of dust storms on the air pollution in Beijing. *Water, Air, and Soil Pollution: Focus, 3*(2), 93–101.

Fann, N., & Risley, D. (2013). The public health context for PM 2.5, and ozone air quality trends. *Air Quality, Atmosphere and Health, 6*(1), 1–11.

Files, D. S., Webb, J. T., & Pilmanis, A. A. (2005). Depressurization in military aircraft: Rates, rapidity, and health effects for 1055 incidents. *Aviation Space and Environmental Medicine, 76*(76), 523–529.

Filonchyk, M., Yan, H., Yang, S., & Hurynovich, V. (2016). A study of PM2.5 and PM10 concentrations in the atmosphere of large cities in Gansu Province, China, in summer period. *Journal of Earth System Science, 125*(6), 1175–1187.

Filonchyk, M., Yan, H., & Hurynovich, V. (2017). Temporal-spatial variations of air pollutants in Lanzhou, Gansu Province, China, during the spring–summer periods, 2014–2016. *Environmental Quality Management, 26*(4), 65–74.

Forsberg, B., Stjernberg, N., Falk, M., Lundbäck, B., & Wall, S. (1993). Air pollution levels, meteorological conditions and asthma symptoms. *European Respiratory Journal, 6*(8), 1109–1115.

Fraser, R. S. (1976). Satellite measurement of mass of Sahara dust in the atmosphere. *Applied Optics, 15*(10), 2471–2479.

Fraser, R. S., Kaufman, Y. J., & Mahoney, R. L. (1984). Satellite measurements of aerosol mass and transport. *Atmospheric Environment, 18*(12), 2577–2584.

Gartland, L. M. (2012). *Heat islands: Understanding and mitigating heat in urban areas*. London: Routledge.

Gillette, D. A., Blifford, I. H. J., & Fenster, C. R. (2010). Measurements of aerosol size distributions and vertical fluxes of aerosols on land subject to wind erosion. *Journal of Applied Meteorology, 11*(6), 977–987.

Giovannini, M., Sala, M., Riva, E., & Radaelli, G. (2010). Hospital admissions for respiratory conditions in children and outdoor air pollution in Southwest Milan, Italy. *Acta Paediatrica, 99*(8), 1180–1185.

Gobbi, G. P., Barnaba, F., & Ammannato, L. (2007). Estimating the impact of Saharan dust on the year 2001 PM 10, record of Rome, Italy. *Atmospheric Environment, 41*(2), 261–275.

Griggs, M. (1975). Measurements of atmospheric aerosol optical thickness over water using ERTS-1 data. *Journal of the Air and Waste Management Association, 25*(6), 622.

Gu, J., Yu, S. S., Zhou, J. L., Zheng, J., & Chen, J. (2001). DCT coefficient and error concealment. *High Technology Letters, 11*(7), 36–39. (in Chinese).

Guo, J. P., Zhang, X. Y., Che, H. Z., Gong, S. L., An, X., & Cao, C. X. (2009). Correlation between PM concentrations and aerosol optical depth in eastern China. *Atmospheric Environment, 43*(37), 5876–5886.

Guo, J., Deng, M., Fan, J., Li, Z., Chen, Q., & Zhai, P. (2014). Precipitation and air pollution at mountain and plain stations in northern china: Insights gained from observations and modeling. *Journal of Geophysical Research Atmospheres, 119*(8), 4793–4807.

Gupta, P., Christopher, S. A., Wang, J., Gehrig, R., Lee, Y., & Kumar, N. (2006). Satellite remote sensing of particulate matter and air quality assessment over global cities. *Atmospheric Environment, 40*(30), 5880–5892.

Hadjimitsis, D. G., Clayton, C. R. I., & Hope, V. S. (2004). An assessment of the effectiveness of atmospheric correction algorithms through the remote sensing of some reservoirs. *International Journal of Remote Sensing, 25*(18), 3651–3674.

Hadjimitsis, D. G., Nisantzi, A., & Trigkas, V. (2010). Satellite remote sensing, GIS and sun-photometers for monitoring PM10 in Cyprus: Issues on public health. *Proceedings of SPIE - The International Society for Optical Engineering, 7826*(20), 78262C-9.

Han, F. G., Zhao, M., Chang, Z. F., Han, S. H., Zhong, S. N., & Guo, S. J. (2009). Analysis on dust weather and its affecting factors in Minqin Sandy Area in recent 50 years: Analysis on dust weather and its affecting factors in Minqin Sandy Area in recent 50 years. *Arid Zone Research, 26*(6), 889–894.

He, X. W., Xue, Y., Li, Y. J., Guang, J., Yang, L. K., & Xu, H. (2012). Monitoring the Haze using multi-sensor aerosol optical depth data. *Egu General Assembly, 14*, 13754.

Herman, B. M., Browning, S. R., & Curran, R. J. (1971). The effect of atmospheric aerosols on scattered sunlight. *Journal of the Atmospheric Sciences, 28*(3), 419–428.

Herman, J. R., Bhartia, P. K., Torres, O., Hsu, C., Seftor, C., & Celarier, E. (1997). Global distribution of UV-absorbing aerosols from Nimbus 7/TOMS data. *Journal of Geophysical Research Atmospheres, 1021*(D14), 16911–16922.

Hien, P. D., Bac, V. T., Tham, H. C., Nhan, D. D., & Vinh, L. D. (2002). Influence of meteorological conditions on PM2.5 and PM2.5−10 concentrations during the monsoon season in Hanoi, Vietnam. *Atmospheric Environment, 36*(21), 3473–3484.

Hogan, T. F., & Brody, L. R. (1993). Sensitivity studies of the Navy's global forecast model parameterizations and evaluation of improvements to NOGAPS. *Monthly Weather Review, 121*(8), 2373–2395.

Holben, B., Vermote, E., Kaufman, Y. J., Tanre, D., & Kalb, V. (1992). Aerosol retrieval over land from AVHRR data-application for atmospheric correction. *IEEE Transactions on Geoscience and Remote Sensing, 30*(2), 212–222.

Holmes, N. S., & Morawska, L. (2006). A review of dispersion modelling and its application to the dispersion of particles: An overview of different dispersion models available. *Atmospheric Environment, 40*(30), 5902–5928.

Hsu, N. C., Tsay, S. C., King, M. D., & Herman, J. R. (2004). Aerosol properties over bright-reflecting source regions. *IEEE Transactions on Geoscience and Remote Sensing, 42*(3), 557–569.

Hsu, N. C., Tsay, S. C., King, M. D., & Herman, J. R. (2006). Deep blue retrievals of Asian aerosol properties during ACE-Asia. *IEEE Transactions on Geoscience and Remote Sensing, 44*(11), 3180–3195.

Hsu, N. C., Jeong, M. J., Bettenhausen, C., Sayer, A. M., Hansell, R., Seftor, C. S., Huang, J., & Tsay, S. C. (2013). Enhanced deep blue aerosol retrieval algorithm: The second generation. *Journal of Geophysical Research: Atmospheres, 118*, 9296–9315.

Huang, W. W., Yang, J., Ling, S. B., & Niu, S. J. (2007). Effects of the heterogeneous-phase chemical processes on mineral aerosols on the growth of cloud droplets. *Journal of Nanjing Institute of Meteorology, 30*(2), 210–215. (In Chinese).

Hunt, W. H., Winker, D. M., Vaughan, M. A., Powell, K. A., Lucker, P. L., & Weimer, C. (2009). CALIPSO Lidar description and performance assessment. *Journal of Atmospheric and Oceanic Technology, 26*(7), 1214–1228.

Hutchison, K. D., Faruqui, S. J., & Smith, S. (2008). Improving correlations between MODIS aerosol optical thickness and ground-based PM 2.5, observations through 3D spatial analyses. *Atmospheric Environment, 42*(3), 530–543.

Iii, C. A. P., Burnett, R. T., Thun, M. J., Calle, E. E., Krewski, D., & Ito, K. (2002). Lung cancer, cardiopulmonary mortality, and long-term exposure to fine particulate air pollution. *The Journal of the American Medical Association, 287*(9), 1132–1141.

Ipsen, J., Deane, M., & Ingenito, F. E. (1969). Relationships of acute respiratory disease to atmospheric pollution and meteorological conditions. *Archives of Environmental Health, 18*(4), 462–472.

Jiang, D., Wang, S., Lang, X., Shang, K. Z., & Yang, D. B. (2001). The characteristics of stratification of lower-layer atmospheric temperature and their relations with air pollution in Lanzhou Proper. *Journal of Lanzhou University, 37*(4), 133–139. (in Chinese).

Jickells, T. D., An, Z. S., Andersen, K. K., Baker, A. R., Bergametti, G., & Brooks, N. (2005). Global iron connections between desert dust, ocean biogeochemistry, and climate. *Science, 308*(5718), 67–71.

Julien, Y., & Sobrino, J. A. (2009). The Yearly Land Cover Dynamics (YLCD) method: An analysis of global vegetation from NDVI and LST parameters. *Remote Sensing of Environment, 113*(2), 329–334.

Jun, L. I., Sun, C. B., Liu, X. D., Dong, S. P., Guo, J., & Wang, Y. (2009). Non-parameter statistical analysis of impacts of meteorological conditions on PM concentration in Beijing. *Research of Environmental Sciences, 22*(6), 663–669. (in Chinese).

Kahn, R., Banerjee, P., & McDonald, D. (2001). Sensitivity of multiangle imaging to natural mixtures of aerosols over ocean. *Journal of Geophysical Research: Atmospheres, 106*(D16), 18219–18238.

Kalashnikova, O. V., & Kahn, R. A. (2008). Mineral dust plume evolution over the Atlantic from MISR and MODIS aerosol retrievals. *Journal of Geophysical Research: Atmospheres, 113*, D24204.

Kan, H., London, S. J., Chen, G., Zhang, Y., Song, G., & Zhao, N. (2008). Season, sex, age, and education as modifiers of the effects of outdoor air pollution on daily mortality in Shanghai, China: The Public Health and Air Pollution in Asia (PAPA) Study. *Environmental Health Perspectives, 116*(9), 1183.

Karlsson, L., Hernandez, F., Rodríguez, S., López-Pérez, M., Hernandez-Armas, J., & Alonso-Pérez, S. (2008). Using 137Cs and 40K to identify natural Saharan dust contributions to PM10 concentrations and air quality impairment in the Canary Islands. *Atmospheric Environment, 42*(30), 7034–7042.

Kaskaoutis, D. G., Rashki, A., Houssos, E. E., Mofidi, A., Goto, D., & Bartzokas, A. (2014). Meteorological aspects associated with dust storms in the Sistan region, southeastern Iran. *Climate Dynamics, 45*(1-2), 407–424.

Kassteele, J. V. D., Koelemeijer, R. B. A., Dekkers, A. L. M., Schaap, M., Homan, C. D., & Stein, A. (2006). Statistical mapping of PM10 concentrations over Western Europe using secondary information from dispersion modeling and MODIS satellite observations. *Stochastic Environmental Research and Risk Assessment, 21*(2), 183–194.

Kaufman, Y. J., & Joseph, J. H. (1982). Determination of surface albedos and aerosol extinction characteristics from satellite imagery. *Journal of Geophysical Research: Oceans, 87*(C2), 1287–1299.

Kaufman, Y. J., & Remer, L. A. (1994). Detection of forests using mid-IR reflectance: An application for aerosol studies. *IEEE Transactions on Geoscience and Remote Sensing, 32*(3), 672–683.

Kaufman, Y. J., & Sendra, C. (1988). Algorithm for automatic atmospheric corrections to visible and near-IR satellite imagery. *International Journal of Remote Sensing, 9*(8), 1357–1381.

Kaufman, Y. J., Koren, I., Remer, L. A., Rosenfeld, D., Rudich, Y., & Ramanathan, V. (2005). The effect of smoke, dust, and pollution aerosol on shallow cloud development over the Atlantic Ocean. *Proceedings of the National Academy of Sciences of the United States of America, 102*(32), 11207–11212.

Khan, S., Cao, Q., Zheng, Y. M., Huang, Y. Z., & Zhu, Y. G. (2008). Health risks of heavy metals in contaminated soils and food crops irrigated with wastewater in Beijing, China. *Environmental Pollution, 152*(3), 686–692.

Kim, K. H., Kabir, E., & Kabir, S. (2015). A review on the human health impact of airborne particulate matter. *Environment International, 74*, 136–143.

Kotarba, A. Z. (2016). Regional high-resolution cloud climatology based on MODIS cloud detection data. *International Journal of Climatology, 36*(8), 3105–3115.

Kuo, Y. M., Chiu, C. H., & Yu, H. L. (2015). Influences of ambient air pollutants and meteorological conditions on ozone variations in Kaohsiung, Taiwan. *Stochastic Environmental Research and Risk Assessment, 29*(3), 1037–1050.

Kurita, H., Sasaki, K., Muroga, H., Ueda, H., & Wakamatsu, S. (2010). Long-range transport of air pollution under light gradient wind conditions. *Journal of Applied Meteorology, 24*(5), 425–434.

Laña, I., Ser, J. D., Padró, A., Vélez, M., & Casanova-Mateo, C. (2016). The role of local urban traffic and meteorological conditions in air pollution: A data-based case study in Madrid. *Spain, Atmospheric Environment, 145*, 424–438.

Lanzhou Meteorological Administration (LMA). (2013). *The weather report data in Lanzhou*. Retrieved from http://qxj.lanzhou.gov.cn/

Lazzarini, M., Marpu, P. R., & Ghedira, H. (2013). Temperature-land cover interactions: The inversion of urban heat island phenomenon in desert city areas. *Remote Sensing of Environment, 130*(4), 136–152.

Lee, J. T., Kim, H., Hong, Y. C., & Kwon, H. J. (2000). Air pollution and daily mortality in seven major cities of Korea, 1991-1997. *Environmental Research, 84*(3), 247–254.

Lee, H. J., Liu, Y., Coull, B. A., Schwartz, J., & Koutrakis, P. (2011a). A novel calibration approach of MODIS AOD data to predict PM2.5 concentrations. *Atmospheric Chemistry and Physics, 11*(11), 9769–9795.

Lee, H. J., Liu, Y., Coull, B. A., Schwartz, J., & Koutrakis, P. (2011b). A novel calibration approach of MODIS AOD data to predict PM2.5 concentrations. *Atmospheric Chemistry and Physics, 11*(11), 9769–9795.

Lee, H. J., Chatfield, R. B., & Strawa, A. W. (2016). Enhancing the applicability of satellite remote sensing for PM2.5 estimation using MODIS Deep Blue AOD and Land Use Regression in California, United States. *Environmental Science and Technology, 50*(12), 6546–6555.

Legrand, M., Desbois, M., & Vovor, K. (1988). Satellite detection of Saharan dust: Optimized imaging during nighttime. *Journal of Climate, 1*(3), 256–264.

Leung, A. O., Duzgorenaydin, N. S., Cheung, K. C., & Wong, M. H. (2008). Heavy metals concentrations of surface dust from e-waste recycling and its human health implications in southeast China. *Environmental Science and Technology, 42*(7), 2674–2680.

Levy, R. C., Remer, L. A., Kleidman, R. G., & Mattoo, S. (2010a). Global evaluation of the Collection 5 MODIS dark-target aerosol products over land. *Atmospheric Chemistry and Physics & Discussions, 10*(6), 10399–10420.

Levy, R. C., Remer, L. A., Kleidman, R. G., & Mattoo, S. (2010b). Global evaluation of the Collection 5 MODIS dark-target aerosol products over land. *Atmospheric Chemistry and Physics and Discussions, 10*(6), 10399–10420.

Levy, R. C., Mattoo, S., Munchak, L. A., & Remer, L. A. (2013). The Collection 6 MODIS aerosol products over land and ocean. *Atmospheric Measurement Techniques, 6*(11), 2989–3034.

Li, X. (2003). Retrieval method for optical thickness of aerosds over Beijing and its vicinity by using the Modis data. *Acta Meteorologica Sinica, 61*(5), 580–591. (in Chinese).

Li, X., & Cheng, H. (2013). Influence of dust weather on air quality in Lanzhou City. *Meteorological and Environmental Research, 11*, 52–54. (in Chinese).

Li, C., Mao, J., & Lau, K. H. (2003). Research on air pollution in Beijing and its surroundings with MODIS aerosol products. *Chinese Journal of Atmospheric Sciences, 27*(5), 419–430.

Li, C. C., Mao, J. T., & Lau, A. K. (2005). Remote sensing of high spatial resolution aerosol optical depth with MODIS data over Hong Kong. *Chinese Journal of Atmospheric Sciences, 29*(3), 324–342. (in Chinese).

Li, Z., Wang, Y., Zhou, Q., Wu, J., Peng, J., & Chang, H. (2008). Spatiotemporal variability of land surface moisture based on vegetation and temperature characteristics in Northern Shaanxi Loess Plateau, China. *Journal of Arid Environments, 72*(6), 974–985.

Li, N., Peng, X. W., & Zhang, B. Y. (2009). Relationship between air pollutant and daily hospital visits for respiratory diseases in Guangzhou: A time-series study. *Journal of Environment and Health, 26*(12), 1077–1080. (in Chinese).

Liu, L., & Zhang, Y. (2011). Urban heat island analysis using the Landsat TM Data and ASTER Data: A case study in Hong Kong. *Remote Sensing, 3*(7), 1535–1552.

Liu, Y., Park, R. J., Jacob, D. J., Li, Q., Kilaru, V., & Sarnat, J. A. (2004a). Mapping annual mean ground-level PM2.5 concentrations using. *Multiangle Imaging Spectroradiometer aerosol optical thickness over the contiguous United States, Journal of Geophysical Research Atmospheres, 109*(D22), 2285–2311.

Liu, Y., Sarnat, J. A., Coull, B. A., Koutrakis, P., & Jacob, D. J. (2004b). Validation of Multiangle Imaging Spectroradiometer (MISR) aerosol optical thickness measurements using Aerosol Robotic Network (AERONET) observations over the contiguous United States. *Journal of Geophysical Research Atmospheres, 109*(D6), 127–128.

Liu, Y., Franklin, M., Kahn, R., & Koutrakis, P. (2007). Using aerosol optical thickness to predict ground-level PM 2.5, concentrations in the St. Louis area: A comparison between MISR and MODIS. *Remote Sensing of Environment, 107*(1–2), 33–44.

Liu, S. M., Yang, H., Zhao, F. U., & Shao, Z. H. (2008). Analysis of meteorological condition and related heavy pollution of PM_(10) in Lanzhou City during winter and spring. *Environmental Science and Technology, 31*(8), 80–83. (in Chinese).

Liu, H., Zhang, X., Yingqui, L. I., Guo, H., & MAIMAITI. (2009). The time-space distribution characteristics of sandstorm weather in Hotan region. *Journal of Arid Land Resources and Environment, 23*(5), 85–89. (in Chinese).

Lu, D., & Weng, Q. (2006). Spectral mixture analysis of ASTER images for examining the relationship between urban thermal features and biophysical descriptors in Indianapolis, Indiana, USA. *Remote Sensing of Environment, 104*(2), 157–167.

Lu, D., Song, K., Zang, S., Jia, M., Du, J., & Ren, C. (2015). The effect of urban expansion on urban surface temperature in Shenyang, China: An analysis with Landsat Imagery. *Environmental Modeling and Assessment, 20*(3), 197–210.

Ma, J. Z., Ding, Z., Gates, J. B., & Su, Y. (2008). Chloride and the environmental isotopes as the indicators of the groundwater recharge in the Gobi Desert, northwest China. *Environmental Geology, 55*(7), 1407–1419.

Mallone, S., Stafoggia, M., Faustini, A., Gobbi, G. P., Marconi, A., & Forastiere, F. (2011). Saharan dust and associations between particulate matter and daily mortality in Rome, Italy. *Environmental Health Perspectives, 119*(10), 1409–1414.

Mao, J., Li, C., Zhang, J., & Lau, K. H. (2002). The comparison of remote sensing aerosol optical depth from MODIS data and Ground Sun-Photometer observation. *Journal of Applied Meteorological Science, 13*(s1), 127–135. (in Chinese).

Martonchik, J. V. (1997). Determination of aerosol optical depth and land surface directional reflectances using multiangle imagery. *Journal of Geophysical Research Atmospheres, 102*(D14), 17015–17022.

Martonchik, J. V., & Diner, D. J. (1992). Retrieval of aerosol optical properties from multi-angle satellite imagery. *IEEE Transactions on Geoscience and Remote Sensing, 30*(2), 223–230.

Martonchik, J. V., Diner, D. J., Kahn, R. A., Ackerman, T. P., Verstraete, M. M., & Pinty, B. (1998). Techniques for the retrieval of aerosol properties over land and ocean using multiangle imaging. *IEEE Transactions on Geoscience and Remote Sensing, 36*(4), 1212–1227.

Masson, O., Piga, D., Gurriaran, R., & D'Amico, D. (2010). Impact of an exceptional Saharan dust outbreak in France: PM10 and artificial radionuclides concentrations in air and in dust deposit. *Atmospheric Environment, 44*(20), 2478–2486.

Matsui, T., Masunaga, H., Sr, R. A. P., & Tao, W. K. (2004). Impact of aerosols and atmospheric thermodynamics on cloud properties within the climate system. *Geophysical Research Letters, 31*(6), 315–336.

Mayer, H. (1999). Air pollution in cities. *Atmospheric Environment, 33*(24–25), 4029–4037.

Mcmurry, P. H., Shepherd, M. F., & Vickery, J. S. (2004). *Particulate matter science for policy makers: A NARSTO assessment*. Cambridge: Cambridge University Press.

Medinaramón, M., Zanobetti, A., & Schwartz, J. (2006). The effect of ozone and PM10 on hospital admissions for pneumonia and chronic obstructive pulmonary disease: A national multicity study. *American Journal of Epidemiology, 163*(6), 579–588.

Meng, Z. Q., Zhang, J., Geng, H., Lu, B., & Zhang, Q. X. (2007). Influence of dust storms on daily respiratory and circulatory outpatient number. *China Environmental Science, 6*(4), 340–357.

Oguz, H. (2013). LST calculator: A program for retrieving land surface temperature from Landsat TM/ETM+ imagery. *Environmental Engineering and Management Journal, 12*(3), 549–555.

Okuyama, K., Shimada, M., Choi, M., & Han, B. (2005). Aerosol particle classification apparatus. *US, US, 20050180543*, A1.

Perevedentsev, Y. P., & Khabutdinov, Y. G. (2012). Meteorological potential of natural self-purification and air quality in Kazan in last decades. Bulletin of Udmurt University, 3, 23–28. (in Russian).

Perez-Padilla, R., Schilmann, A., & Riojas-Rodriguez, H. (2010). Respiratory health effects of indoor air pollution Review article. *The International Journal of Tuberculosis and Lung Disease, 14*(9), 1079–1086.

Peterson, T. C. (2003). Assessment of urban versus rural in situ surface temperatures in the contiguous united states: No difference found. *Journal of Climate, 16*(18), 2941–2959.

Pope, C. A., & Dockery, D. W. (2006). Health effects of fine particulate air pollution: Lines that connect. *Journal of the Air & Waste Management Association, 56*(6), 709–742.

Pope, P. C., Bates, D. V., & Raizenne, M. E. (1995). Health effects of particulate air pollution: Time for reassessment? *Environmental Health Perspectives, 103*(5), 472–480.

Pope, C. A., Hill, R. W., & Villegas, G. M. (1999). Particulate air pollution and daily mortality on Utah's Wasatch Front. *Environmental Health Perspectives, 107*(7), 567–573.

Pu, R., Gong, P., Michishita, R., & Sasagawa, T. (2006). Assessment of multi-resolution and multi-sensor data for urban surface temperature retrieval. *Remote Sensing of Environment, 104*(2), 211–225.

Qi, W., Liao, Y., & Yi, M. (2011). A pilot study on health-based assessment for economic loss caused by sand-dust weather in China. *Journal of Environment and Health, 28*(9), 804–808.

Qian, W., Tang, X., & Quan, L. (2004). Regional characteristics of dust storms in China. *Atmospheric Environment, 38*(29), 4895–4907.

Qiang, M., Chen, F., Zhou, A., Xiao, S., Zhang, J., & Wang, Z. (2007). Impacts of wind velocity on sand and dust deposition during dust storm as inferred from a series of observations in the northeastern Qinghai–Tibetan Plateau, China. *Powder Technology, 175*(2), 82–89.

Qiao, Z., Tian, G., & Xiao, L. (2013). Diurnal and seasonal impacts of urbanization on the urban thermal environment: A case study of Beijing using MODIS data. *ISPRS Journal of Photogrammetry and Remote Sensing, 85*(2), 93–101.

Qiu, J. (1995). A new method of determining atmospheric aerosol optical depth from the whole-spectral solar direct radiation, Part I: Theory. *Scientia Atmospherica Sinica, 9*(5), 385–394. (in Chinese).

Remer, L. A., Kaufman, Y. J., Tanré, D., Mattoo, S., Chu, D. A., & Martins, J. V. (2005). The MODIS aerosol algorithm, products, and validation. *Journal of Atmospheric Sciences, 62*(4), 947–973.

Remoundaki, E., Bourliva, A., Kokkalis, P., Mamouri, R. E., Papayannis, A., & Grigoratos, T. (2011). PM10 composition during an intense Saharan dust transport event over Athens (Greece). *Science of the Total Environment, 409*(20), 4361–4372.

Retalis, A., Hadjimitsis, D. G., Michaelides, S., & Tymvios, F. (2010). Comparison of aerosol optical thickness with in situ visibility data over Cyprus. *Natural Hazards and Earth System Sciences, 10*(3), 421–428.

Reuter, D. C., Richardson, C. M., Pellerano, F. A., Irons, J. R., Allen, R. G., & Anderson, M. (2015). The Thermal Infrared Sensor (TIRS) on Landsat 8: Design overview and pre-launch characterization. *Remote Sensing, 7*(1), 1135–1153.

Rolph, G., Stein, A., & Stunder, B. (2017). Real-time environmental applications and display system: READY. *Environmental Modelling and Software, 95*, 210–228.

Rosenfeld, D., Dai, J., Yu, X., Yao, Z., Xu, X., & Yang, X. (2007). Inverse relations between amounts of air pollution and orographic precipitation. *Science, 315*(5817), 1396–1398.

Roy, D., Gautam, S., Singh, P., Singh, G., Das, B. K., & Patra, A. K. (2016). Carbonaceous species and physicochemical characteristics of PM 10, in coal mine fire area—A case study. *Air Quality, Atmosphere and Health, 9*(4), 429–437.

Rutllant, J., & Garreaud, R. (1995). Meteorological air pollution potential for Santiago, Chile: Towards an objective episode forecasting. *Environmental Monitoring and Assessment, 34*(3), 223.

Samoli, E., Peng, R., Ramsay, T., Pipikou, M., Touloumi, G., & Dominici, F. (2008). Acute effects of ambient particulate matter on mortality in Europe and North America: Results from the APHENA study. *Environmental Health Perspectives, 116*(11), 1480.

Sayer, A. M., Munchak, L. A., Hsu, N. C., Levy, R. C., Bettenhausen, C., & Jeong, M. (2015a). MODIS Collection 6 aerosol products: Comparison between Aqua's e-Deep Blue, Dark Target, and "merged" data sets, and usage recommendations. *Journal of Geophysical Research Atmospheres, 119*(24), 13965–13989.

Sayer, A. M., Hsu, N. C., Bettenhausen, C., Jeong, M. J., & Meister, G. (2015b). Effect of MODIS Terra radiometric calibration improvements on Collection 6 Deep Blue aerosol products: Validation and Terra/Aqua consistency. *Journal of Geophysical Research, 120*(23), 12157–12174.

Schindler, C., Keidel, D., Gerbase, M. W., Zemp, E., Bettschart, R., & Brändli, O. (2009). Improvements in PM10 exposure and reduced rates of respiratory symptoms in a cohort of Swiss adults (SAPALDIA). *American Journal of Respiratory and Critical Care Medicine, 179*(7), 579–587.

Schmidt-Ott, A., & Büscher, P. (1991). In situ chemical classification of atmospheric aerosol particles. *Journal of Aerosol Science, 22*(6), S307.

Schwartz, J. (1993). Air pollution and daily mortality in Birmingham, Alabama. *American Journal of Epidemiology, 137*(10), 1136–1147.

Seleguei, T. S. (1989). Meteorological potential of atmosphere self-cleaning of the Siberian economic region. *Trudy Zapadno-Sibirskogo regional'nogo NII Goskomgidrometa, 86*, 84–89. (in Russian).

Seleguei, T. S., & Yurchenko, P. I. (1990). Potential scattering power day. *Geography and Natural Resources, Novosibirsk: Science, 2*, 132–137. (in Russian).

Shi, T., Yang, Y., Ma, J., Zhang, L., & Luo, S. (2013). Spatial-temporal characteristics of urban heat Island in typical cities of Anhui Province Based on MODIS. *Journal of Applied Meteorological Science, 24*(4), 484–494. (in Chinese).

Shiraiwa, M., Selzle, K., & Pöschl, U. (2012). Hazardous components and health effects of atmospheric aerosol particles: Reactive oxygen species, soot, polycyclic aromatic compounds and allergenic proteins. *Free Radical Reserch, 46*(8), 927–939.

Slater, J. F., Dibb, J. E., Campbell, J. W., & Moore, T. S. (2004). Physical and chemical properties of surface and column aerosols at a rural New England site during MODIS overpass. *Remote Sensing of Environment, 92*(2), 173–180.

Sobrino, J. A., Jiménez-Muñoz, J. C., & Paolini, L. (2004). Land surface temperature retrieval from LANDSAT TM 5. *Remote Sensing of Environment, 90*(4), 434–440.

Song, C., Jia, L., & Menenti, M. (2014). Retrieving high-resolution surface soil moisture by downscaling AMSR-E brightness temperature using MODIS LST and NDVI Data. *IEEE Journal of Selected Topics in Applied Earth Observations and Remote Sensing, 7*(3), 935–942.

Sonkin, L. R., & Nikolaev, V. D. (1993). Synoptic analysis and atmospheric pollution forecast. *Russian Meteorology and Hydrology, 5*, 10–14.

Sonkin, L. R., Nikolaev, V. D., Ivleva, T. P., & Kirillova, V. I. (2002). Forecasts of extremely high levels of air pollution in cities and regions. In S. S. Chicherin (Ed.), *Problems of atmospheric boundary-layer physics and air pollution* (pp. 310–322). Hydrometeorological Publishers, Saint Petersburg.

Srivastava, P. K., Majumdar, T. J., & Bhattacharya, A. K. (2009). Surface temperature estimation in Singhbhum Shear Zone of India using Landsat-7 ETM+ thermal infrared data. *Advances in Space Research, 43*(10), 1563–1574.

Stein, A. F., Draxler, R. R., Rolph, G. D., Stunder, B. J. B., Cohen, M. D., & Ngan, F. (2016). NOAA's HYSPLIT atmospheric transport and dispersion modeling system. *Bulletin of the American Meteorological Society, 96*(12), 2059–2077.

Su, X. (2008). New advances of dust aerosol in China. *Meteorological and Environmental Sciences, 31*(3), 72–77. (in Chinese).

Sun, D., & Kafatos, M. (2007). Note on the NDVI-LST relationship and the use of temperature-related drought indices over North America. *Geophysical Research Letters, 34*(24), 497–507.

Sun, L., Liu, Q. H., Chen, L. F., & Liu, Q. (2006). The application of HJ-1 Hyperspectral Imaging Radiometer to retrieve aerosol optical thickness over land. *Journal of Remote Sensing, 10*(5), 770–776. (in Chinese).

Sun, Y. L., Wang, Z. F., Fu, P. Q., & Yang, T. (2013). Aerosol composition, sources and processes during wintertime in Beijing, China. *Atmospheric Chemistry and Physics Discussions, 13*(9), 4577–4592.

Ta, W., Wang, T., Xiao, H., Zhu, X., & Xiao, Z. (2004). Gaseous and particulate air pollution in the Lanzhou Valley, China. *Science of the Total Environment, 320*(2–3), 163–176.

Ta, W., Wang, H., & Jia, X. (2013). External supply of dust regulates dust emissions from sand deserts. *Catena, 110*(2), 113–118.

Tan, K. C., Lim, H. S., Matjafri, M. Z., & Abdullah, K. (2010). Land surface temperature retrieval by using ATCOR3_T and Normalized Difference Vegetation Index methods in Penang Island. *American Journal of Applied Sciences, 7*(5), 717–723.

Tan, K. C., Lim, H. S., Matjafri, M. Z., & Abdullah, K. (2012). A comparison of radiometric correction techniques in the valuation of the relationship between LST and NDVI in Landsat imagery. *Environmental Monitoring and Assessment, 184*(6), 3813–3829.

Tanner, P. A., & Law, P. T. (2002). Effects of synoptic weather systems upon the air quality in an Asian megacity. *Water, Air, and Soil Pollution, 136*(1), 105–124.

Tanré, D., Devaux, C., Herman, M., & Santer, R. (1988a). Radiative properties of desert aerosols by optical ground-based measurements at solar wavelengths. *Journal of Geophysical Research Atmospheres, 93*(D11), 14223–14231.

Tanré, D., Devaux, C., Herman, M., Santer, R., & Gac, J. Y. (1988b). Radiative properties of desert aerosols by optical ground-based measurements at solar wavelengths. *Journal of Geophysical Research: Atmospheres, 93*(D11), 14223–14231.

Tanré, D., Herman, M., & Kaufman, Y. J. (1996). Information on aerosol size distribution contained in solar reflected spectral radiances. *Journal of Geophysical Research: Atmospheres, 101*(D14), 19043–19060.

Tanré, D., Remer, L. A., Kaufman, Y. J., Mattoo, S., Hobbs, P. V., & Livingston, J. M. (1999). Retrieval of aerosol optical thickness and size distribution over ocean from the MODIS airborne simulator during TARFOX. *Journal of Geophysical Research, 104*(D2), 2261–2278.

Tao, J. H., Huang, Y. X., & Lu, D. R. (2007a). Influence of sand-dust activities in Hexi Corridor on PM10 concentration in Lanzhou and its assessment. *Journal of Desert Research, 27*(4), 272–276. (in Chinese).

Tao, J. H., Huang, Y. X., & Deng-Rong, L. U. (2007b). Influence of sand-dust activities in Hexi Corridor on PM10 concentration in Lanzhou and its assessment. *Journal of Desert Research, 27*(4), 672–676. (in Chinese).

Tao, Y., Mi, S., Zhou, S., Wang, S., & Xie, X. (2014). Air pollution and hospital admissions for respiratory diseases in Lanzhou. *China, Environmental Pollution, 185*, 196–201.

Tian, L., Ran-Ying, L. U., Xing, W. T., Wang, L., Wang, X., & Wang, W. (2005). Studies on city ambient air quality in china during 2001-2004. *Journal of Arid Land Resources & Environment, 19*(7), 101–105. (in Chinese).

Tian, G., Qiao, Z., & Xu, X. (2014). Characteristics of particulate matter (PM10) and its relationship with meteorological factors during 2001-2012 in Beijing. *Environmental Pollution, 192*(192), 266–274.

Tie, X., Brasseur, G. P., Zhao, C. S., Granier, C., Massie, S., & Qin, Y. (2006). Chemical characterization of air pollution in Eastern China and the Eastern United States. *Atmospheric Environment, 40*(14), 2607–2625.

Torres, O., Tanskanen, A., Veihelmann, B., Ahn, C., Braak, R., & Bhartia, P. K. (2007a). Aerosols and surface UV products from ztions: An overview. *Journal of Geophysical Research Atmospheres, 112*(D24), 1–14.

Torres, O., Tanskanen, A., Veihelmann, B., Ahn, C., Braak, R., & Bhartia, P. K. (2007b). Aerosols and surface UV products from Ozone Monitoring Instrument observations: An overview. *Journal of Geophysical Research Atmospheres, 112*(D24), 1–14.

Torres, O., Ahn, C., & Chen, Z. (2013). Improvements to the OMI near UV aerosol algorithm using A-train CALIOP and AIRS observations. *Atmospheric Measurement Techniques, 6*(11), 3257–3270.

Tsai, S. S., Chang, C. C., & Yang, C. Y. (2013). Fine particulate air pollution and hospital admissions for chronic obstructive pulmonary disease: A case-crossover study in Taipe. *International Journal of Environmental Research and Public Health, 10*(11), 6015–6026.

Tsangari, H., Paschalidou, A. K., Kassomenos, A. P., Vardoulakis, S., Heaviside, C., & Georgiou, K. E. (2016). Extreme weather and air pollution effects on cardiovascular and respiratory hospital admissions in Cyprus. *Science of the Total Environment, 542*(Pt A), 247.

U.S. Environmental Protection Agency (US EPA). (2010). *Module 3: Characteristics of particles – Aerodynamic diameter*. Retrieved November 12, 2011, from http://www.epa.gov/apti/bces/module3/diameter/diameter.htm

Van der Zee, S., Hoek, G., Boezen, H. M., Schouten, J. P., Wijnen, J. H. V., & Brunekreef, B. (1999). Acute effects of urban air pollution on respiratory health of children with and without chronic respiratory symptoms. *Occupational and Environmental Medicine, 56*(12), 802–812.

Vardoulakis, S., Gonzalez-Flesca, N., & Fisher, B. E. A. (2002). Assessment of traffic-related air pollution in two street canyons in Paris: Implications for exposure studies. *Atmospheric Environment, 36*(6), 1025–1039.

Veefkind, J. P., Leeuw, G. D., & Durkee, P. A. (1998). Retrieval of aerosol optical depth over land using twoangle view satellite radiometry during TARFOX. *Geophysical Research Letters, 25*(16), 3135–3138.

Vinnikov, K. Y., Yu, Y., Raja, M. K. R. V., Dan, T., & Goldberg, M. D. (2008). Diurnal-seasonal and weather-related variations of land surface temperature observed from geostationary satellites. *Geophysical Research Letters, 35*(22), 113–130.

Voogt, J. A., & Oke, T. R. (2003). Thermal remote sensing of urban climates. *Remote Sensing of Environment, 86*(3), 370–384.

Wan, Z., & Li, Z. L. (1997). A physics-based algorithm for retrieving land-surface emissivity and temperature from EOS/MODIS data. *IEEE Transactions on Geoscience and Remote Sensing, 35*(4), 980–996.

Wang, A. (1999). Recent trends on study of atmospheric aerosols. *Environmental Chemistry, 1*, 10–15.

Wang, P. L. (2005). The study progress in the research for the particular in city air and its effect on human health. *Environmental Monitoring in China, 21*(1), 83–87. (in Chinese).

Wang, J., & Christopher, S. A. (2003). Intercomparison between satellite-derived aerosol optical thickness and PM2.5 mass: Implications for air quality studies. *Geophysical Research Letters, 30*(21), 267–283.

Wang, S., Yang, M., Bin, Q. I., Xin, C. L., & Yang, M. F. (1999a). Influence of sand-dust storms occurring over the Gansu Hexi District on the air pollution in Lanzhou City. *Journal of Desert Research, 19*(4), 354–358. (in Chinese).

Wang, S., Zhang, L., Chen, C., & Yuan, J. (1999b). Retrospect and prospect for the studies of atmospheric environment in the Lanzhou area. *Journal of Lanzhou University, 3*, 189–201. (in Chinese).

Wang, H. W., Lin, G., & Pan, X. D. (2003a). Association between total suspended particles (TSP) and cardiovascular disease mortality in Shenyang. *Journal of Environment and Health, 20*(1), 13–15. (in Chinese).

Wang, X., Yang, S., Zhu, Y., & Yi, W. (2003b). Aerosol optical thickness retrieval over land From MODIS Data based on the inversion of the 6S Model. *Chinese Journal of Quantum Electronics, 20*(5), 629–634. (in Chinese).

Wang, X., Dong, Z., Zhang, J., & Liu, L. (2004). Modern dust storms in China: An overview. *Journal of Arid Environments, 58*(4), 559–574.
Wang, S., Wang, J., Zhou, Z., & Shang, K. (2005). Regional characteristics of three kinds of dust storm events in China. *Atmospheric Environment, 39*(3), 509–520.
Wang, S., Yuan, W., & Shang, K. (2006). The impacts of different kinds of dust events on PM 10, pollution in northern China. *Atmospheric Environment, 40*(40), 7975–7982.
Wang, Z. T., Chen, L. F., & Zhang, Y. (2008). Urban surface aerosol monitoring using DDV method from MODIS data. *Remote Sensing Technology and Application, 23*(3), 284–288. (in Chinese).
Wang, R. D., Zou, X. Y., Cheng, H., & Xiao-Xu, W. U. (2009a). Spatiotemporal characteristics of sand-dust weather and its influence factors in Hebei Province. *Bulletin of Soil and Water Conservation, 2009*(6), 57–63.
Wang, Z. T., Li, Q., Tao, J. H., Li, S. S., Wang, Q., & Chen, L. F. (2009b). Monitoring of aerosol optical depth over land surface using CCD camera on HJ-1 satellite. *China Environmental Science, 29*(9), 902–907. (in Chinese).
Wang, J., Xu, X., Spurr, R., Wang, Y., & Drury, E. (2010). Improved algorithm for MODIS satellite retrievals of aerosol optical thickness over land in dusty atmosphere: Implications for air quality monitoring in China. *Remote Sensing of Environment, 114*(11), 2575–2583.
Wang, X., Zhang, C., Wang, H., Qian, G., Luo, W., & Lu, J. (2011). The significance of Gobi desert surfaces for dust emissions in China: An experimental study. *Environmental Earth Sciences, 64*(4), 1039–1050.
Wang, X., Hua, T., Zhang, C., Lang, L., & Wang, H. (2012). Aeolian salts in Gobi deserts of the western region of Inner Mongolia: Gone with the dust aerosols. *Atmospheric Research, 118*(3), 1–9.
Wang, J., Hu, Z., Chen, Y., Chen, Z., & Xu, S. (2013a). Contamination characteristics and possible sources of PM10 and PM2.5 in different functional areas of Shanghai, China. *Atmospheric Environment, 68*(2), 221–229.
Wang, T., Yan, C. Z., Song, X., & Li, S. (2013b). Landsat images reveal trends in the Aeolian desertification in a source area for sand and dust storms in China's Alashan plateau (1975–2007). *Land Degradation and Development, 24*(5), 422–429.
Wang, F., Qin, Z., Song, C., Tu, L., Karnieli, A., & Zhao, S. (2015a). An improved Mono-Window Algorithm for land surface temperature retrieval from Landsat 8 thermal infrared sensor data. *Remote Sensing, 7*(4), 4268–4289.
Wang, L., Zhao, Y., Yang, X., Jianmin, M. A., Huang, T., & Gao, H. (2015b). Prediction of air quality in Lanzhou Using Time Series Model and Residual Control Chart. *Plateau Meteorology, 24*(1), 97–103. (in Chinese).
Wang, K., Jiang, S., Wang, J., Zhou, C., Wang, X., & Lee, X. (2017). Comparing the diurnal and seasonal variabilities of atmospheric and surface urban heat islands based on the Beijing urban meteorological network. *Journal of Geophysical Research Atmospheres, 122*(4), 2131–2154.
Wei, F. S., Teng, E. J., Wu, G. P., Hu, W., & Wilson, W. E. (2001). Concentrations and elemental components of PM2. 5, PM10 in ambient air in four large Chinese cities. *Environmental Monitoring in China, 17*(7), 1–6. (in Chinese).
Wei, Y. X., Tong, Y. Q., Yin, Y., & Chen, K. (2009). The variety of main air pollutants concentration and its relationship with meteorological condition in Nanjing city. *Transactions of Atmospheric Sciences, 32*(3), 451–457. (in Chinese).
Weng, Q. (2009). Thermal infrared remote sensing for urban climate and environmental studies: Methods, applications, and trends. *ISPRS Journal of Photogrammetry and Remote Sensing, 64*(4), 335–344.
Weng, Q., & Quattrochi, D. A. (2006a). Thermal remote sensing of urban areas: An introduction to the special issue. *Remote Sensing of Environment, 104*(2), 119–122.
Weng, Q., & Quattrochi, D. A. (2006b). Thermal remote sensing of urban areas: An introduction to the special issue. *Remote Sensing of Environment, 104*(2), 119–122.
Weng, Q., Lu, D., & Schubring, J. (2004). Estimation of land surface temperature–vegetation abundance relationship for urban heat island studies. *Remote Sensing of Environment, 89*(4), 467–483.

Whiteman, C. D., Hoch, S. W., Horel, J. D., & Charland, A. (2014). Relationship between particulate air pollution and meteorological variables in Utah's Salt Lake Valle. *Atmospheric Environment, 94*, 742–753.

WHO (World Health Organization). (2005). *World Health Organization air quality guidelines global update*. Report on a Working Group Meeting, Bonn, Germany, 18–20 October 2005. WHO Regional Office for Europe, E87950.

Wilson, A. M., Wake, C. P., Kelly, T., & Salloway, J. C. (2005). Air pollution, weather, and respiratory emergency room visits in two northern New England cities: An ecological time-series study. *Environmental Research, 97*(3), 312–321.

Winker, D. M., Pelon, J., Coakley, J. A. J., Ackerman, S. A., Charlson, R. J., & Colarco, P. R. (2010). The CALIPSO Mission: A global 3D view of aerosols and clouds. *Bulletin of the American Meteorological Society, 91*(9), 1211–1229.

Wong, T. W., Lau, T. S., Yu, T. S., Neller, A., Wong, S. L., & Tam, W. (1999). Air pollution and hospital admissions for respiratory and cardiovascular diseases in Hong Kong. *Occupational and Environmental Medicine, 56*(10), 679–683.

World Health Organization. (2006). *WHO air quality guidelines for particulate matter, ozone, nitrogen dioxide and sulfur dioxide, summary of risk assessment*. Geneva: World Health Organization.

World Health Organization. (2007). Health relevance of particulate matter from various sources: Report on a WHO workshop, Bonn, Germany 26-27 March 2007. *Imperial College London, 47*(1), 33–40.

World Health Organization. (2015). *Review of evidence on health aspects of air pollution – REVIHAAP Project: Final technical report*. Geneva: World Health Organization.

Wu, D. (2012). Hazy weather research in China in the last decade: A review. *Acta Scientiae Circumstantiae, 32*, 257–269.

Xiao, Z. H., Shao, L. Y., & Zhang, N. (2012). *Physical and chemical characteristics of airborne particles and its effect on health in Lanzhou*. Jiangsu: China University of Mining and Technology Press.

Xu, H., Wang, Y., Wen, T., Yang, Y., & Zhao, Y. (2009). Characteristics and source apportionment of atmospheric aerosols at the summit of Mount Tai during summertime. *Atmospheric Chemistry and Physics, 9*(4), 16361–16379.

Xu, Y., Qin, Z., & Wan, H. (2010). Spatial and temporal dynamics of urban heat island and their relationship with land cover changes in urbanization process: A case study in Suzhou, China. *Journal of the Indian Society of Remote Sensing, 38*(4), 654–663.

Yang, X., & Li, Z. (2014). Increases in thunderstorm activity and relationships with air pollution in southeast China. *Journal of Geophysical Research Atmospheres, 119*(4), 1835–1844.

Yang, J. S., Wang, Y. Q., & August, P. V. (2004). Estimation of land surface temperature using spatial interpolation and satellite-derived surface emissivity. *Journal of Environmental Informatics, 4*(1), 37–44.

Yang, L. X., Wang, D. C., Cheng, S. H., Wang, Z., Zhou, Y., & Zhou, X. H. (2007). Influence of meteorological conditions and particulate matter on visual range impairment in Jinan, China. *Science of the Total Environment, 383*(1-3), 164.

Yi, Y., Yang, Z., & Zhang, S. (2011). Ecological risk assessment of heavy metals in sediment and human health risk assessment of heavy metals in fishes in the middle and lower reaches of the Yangtze River basin. *Environmental Pollution, 159*(10), 2575–2585.

Yin, X. H., Shi, S. Y., Zhang, M. Y., & Li, J. (2007). Change characteristic of Beijing dust weather and its sand-dust source areas. *Plateau Meteorology, 26*(5), 1039–1044. (in Chinese).

Yu, Y., Xia, D. S., Chen, L. H., Liu, N., Chen, J. B., & Gao, Y. H. (2010). Analysis of particulate pollution characteristics and its causes in Lanzhou, Northwest China. *Environmental Science, 31*(1), 22. (in Chinese).

Yu-Lin, L. I., Wan-Quan, T., & Cui, J. Y. (2006). Comparison of particle size between aeolian dusts collected under dust event and non-dust event. *Journal of Desert Research, 26*(4), 644–647. (in Chinese).

Zhang, Y. P., & Jin-Fen, L. I. (2008). Exposure-response relationship between particulate pollution level and hospital outpatient visits in Taiyuan. *Journal of Environment and Health, 25*(6), 479–482. (in Chinese).

Zhang, J., & Tang, C. G. (2012). Vertical distribution structure and characteristic of aerosol over Arid region in a dust process of spring. *Plateau Meteorology, 31*(1), 156–166.

Zhang, X. Y., Arimoto, R., & An, Z. S. (1997). Dust emission from Chinese desert sources linked to variations in atmospheric circulation. *Journal of Geophysical Research Atmospheres, 102*(23), 28041–28047.

Zhang, X. Y., Arimoto, R., Zhu, G. H., Chen, T., & Zhang, G. Y. (1998). Concentration, size-distribution and deposition of mineral aerosol over Chinese desert regions. *Tellus Series B-Chemical and Physical Meteorology, 50*(4), 317–330.

Zhang, J., Si, Z., Mao, J., & Wang, M. (2003). Remote sensing aerosol optical depth over China with GMS-5 Satellite. *Chinese Journal of Atmospheric Sciences, 27*(1), 23–35. (in Chinese).

Zhang, J., Wang, Y., & Li, Y. (2006). A C++ program for retrieving land surface temperature from the data of Landsat TM/ETM+ band6. *Computers and Geosciences, 32*(10), 1796–1805.

Zhang, L., Cao, X., Bao, J., & Zhou, B. (2010). A case study of dust aerosol radiative properties over Lanzhou, China. *Atmospheric Chemistry and Physics, 10*(9), 4283–4293.

Zhang, Z., Wang, J., Chen, L., Chen, X., Sun, G., & Zhong, N. (2014). Impact of haze and air pollution-related hazards on hospital admissions in Guangzhou, China. *Environmental Science and Pollution Research, 21*(6), 4236–4244.

Zhao, T. L., Gong, S. L., Zhang, X. Y., Blanchet, J. P., Mckendry, I. G., & Zhou, Z. J. (2006). A simulated climatology of Asian dust aerosol and its Trans-Pacific transport. Part I: Mean climate and validation. *Journal of Climate, 19*(1), 88–103.

Zhao, S., Qin, Q., Yang, Y., Xiong, Y., & Qiu, G. (2009). Comparison of two split-window methods for retrieving land surface temperature from MODIS data. *Journal of Earth System Science, 118*(4), 345–353.

Zhao, J. G., Wang, S. G., Wang, J. Y., Jian-Rong, B. I., Shi, J. S., & Wang, T. H. (2013a). Analysis of the relationship between pollution in Lanzhou City and ground meteorological factors. *Journal of Lanzhou University, 49*(4), 491–490. (in Chinese).

Zhao, Y., Huang, A., Zhu, X., Zhou, Y., & Huang, Y. (2013b). The impact of the winter North Atlantic Oscillation on the frequency of spring dust storms over Tarim Basin in northwest China in the past half-century. *Environmental Research Letters, 8*(2), 1–5.

Zhen, B. (2010). Characterization and Source Apportionment of PM_(2.5) and PM_(10) in Hangzhou. *Environmental Monitoring in China, 26*(2), 44–48. (in Chinese).

Zheng, S., Zhao, X., Zhang, H., Qisheng, H. E., & Cao, C. (2011). Atmospheric correction on CCD data of HJ-1 satellite and analysis of its effect. *Journal of Remote Sensing, 15*(4), 709–721.

Zheng, Y., Wang, Q., & Liang, Z. (2016). Aerosol retrieval and atmospheric correction of HJ-1 satellite CCD data over land surface of Taihu lake. *Engineering of Surveying & Mapping, 25*(5), 10–15. (in Chinese).

Zhi, Y., Fu, Z., & Shao, Z. (2007). The dust weather and its influence on air quality in Lanzhou city. *Journal of Catastrophology, 22*(1), 77–81.

Zhou, J., Chen, Y. H., Li, J., & Weng, Q. H. (2008). A volume model for urban heat island based on remote sensing imagery and its application: A case study in Beijing. *Journal of Remote Sensing, 12*(5), 734–742.

Zhu, G., & Blumberg, D. G. (2002). Classification using ASTER data and SVM algorithms: The case study of Beer Sheva, Israel. *Remote Sensing of Environment, 80*(2), 233–240.

Index

M. Filonchyk, H. Yan, *Urban Air Pollution Monitoring by Ground-Based Stations and Satellite Data*, https://doi.org/10.1007/978-3-319-78045-0

Printed in the United States
By Bookmasters